Principles of Atmospheric Physics and Chemistry

Principles of Atmospheric Physics and Chemistry

Richard Goody

New York Oxford
Oxford University Press
1995

Oxford University Press

Oxford New York
Athens Auckland Bangkok Bombay
Calcutta Cape Town Dar es Salaam Delhi
Florence Hong Kong Istanbul Karachi
Kuala Lumpur Madras Madrid Melbourne
Mexico City Nairobi Paris Singapore
Taipei Tokyo Toronto

and associated companies in

Berlin Ibadan

Copyright © 1995 by Oxford University Press, Inc.

Published by Oxford University Press, Inc.,
198 Madison Avenue, New York, New York 10016

Oxford is a registered trademark of Oxford University Press

All rights reserved. No part of this publication
may be reproduced, stored in a retrieval system, or transmitted,
in any form or by any means, electronic, mechanical,
photocopying, recording, or otherwise, without the prior
permission of Oxford University Press.

Library of Congress Cataloging-in-Publication Data
Goody, Richard M.
Principles of atmospheric physics and chemistry / Richard Goody.
p. cm.
Includes bibliographical references and index.
ISBN 0-19-509362-3
1. Atmospheric physics. 2. Atmospheric chemistry. I. Title.
QC861.2.G66 1996
551.5—dc20   95-31990
                                                              CIP

ISBN: 0-19-509362-3

1 3 5 7 9 8 6 4 2

Printed in the United States of America
on acid-free paper

Preface

Environmental policy studies increasingly look to the atmospheric sciences for predictions of future atmospheric states. But the climate system is exceptionally complex and, in many respects, the fundamental basis for predictions is insecure. I consider this book to be an elementary contribution to climate studies in the sense that it deals with the underlying principles. It does so at the level of knowledge of a physics or chemistry major, who may be encountering atmospheric science for the first time. I assume that the student has a working knowledge of thermodynamics, electrodynamics, elementary quantum theory, mechanics, wave optics, and a slight acquaintance with reaction kinetics. I have been urged to include reviews of these topics, particularly thermodynamics, which is notoriously neglected in physics concentrations, but continuity of ideas would suffer and space did not permit.

I have derived all important results from first principles for a simple reason. Major theoretical advances in this subject are invariably based upon clear thinking about fundamental issues. Incremental contributions can extend ideas created by others, but many areas of climate research require new ideas that must be developed from first principles. The reader who needs to be convinced of this statement should study any of the major innovators in atmospheric science since World War I: Jeffreys, Chapman, Charney, Richardson, and Lorenz are a few of the more obvious examples.

This book is intended to serve as a graduate level introduction for motivated students without prior knowledge of atmospheric science. Such was the typical student in Earth and Planetary Physics 200, taught in the Division of Applied Sciences at Harvard University. This course was taught in two parts: EPP 200a dealt with atmospheric physics and chemistry, while EPP 200b was concerned with dynamics. I have attempted to cover only the material of EPP 200a; the material of EPP 200b has also been made into an informal textbook by Richard Lindzen that is referred to in this text.

Those of us who taught EPP 200a at Harvard had difficulty finding a suitable textbook. This is not to belittle the many excellent books on atmospheric science that are now available, but for our audience they were usually too specialized or not in the style to which a physics or chemistry major is accustomed. This book is an attempt to fill this gap. It would have been helpful to me when I first encountered atmospheric problems immediately after I completed my bachelor's degree in physics during World War II, and I hope that it will help like-minded students of a later era.

I have included a selection of problems. Some are fairly difficult, and I have tried to indicate the level of difficulty. I have sometimes supplemented these problems with numerical exercises using programs that are routinely used by the research community. It was not feasible to include these programs in this book, but I strongly urge other teachers to set up similar laboratories. The student profits greatly from hands-on experience at a computer terminal.

A number of colleagues have given permission to reproduce figures from their publications: Dr. John Garratt and Kluwer Academic Publishers (Fig. 9.1); Dr. R. Atkinson (Fig. 6.8); Mr. Sheppard Clough (Fig. 3.7); Professor Robert Fleagle (Fig. 9.2); Dr. James Klett (Figs. 8.4, 8.5); Professor R. R. Rogers (Figs. 8.7, 8.8, 8.13); Sir John Houghton (Fig. 2.10); Dr. Keith Browning (Fig. 8.13); Dr. Abraham Oort and the American Institute of Physics (Figs. 2.2, 2.3, C.2); Professor Julius London (Figs. 2.4, 6.2, B.1); Professor Bruce Albrecht (Fig. 8.12); Professor Kuonan Liou (Fig. 8.10); Sir John Mason (Figs. 8.1, 8.2, 8.3, 8.9., 8.14, J.1); Dr. Susan Solomon (Figs. 3.1, 6.3, 6.5, 6.11, 6.12); Baron Marcel Nicolet (Figs. 3.2, 6.4); Dr. Peter Warneck (Figs. 1.6, 6.6, 7.1, 7.2, 7.3, 7.4, 7.5, 7.6, 7.9, C.3, C.4); Professor Michael McElroy (Fig. 6.9); Professor R. E. Newell (Figs. B.2, B.3). I wish to express my thanks to all of them.

On a more personal level, I am particularly grateful to Professor Yuk Ling Yung, Professor Mario Molina, Dr. Ralph Kahn, and Mr. Albert Yen for many helpful comments on the manuscript and for the discovery of many errors, and to Professor Richard Lindzen for encouragement.

Falmouth, Mass.   R.G.
June, 1995

Contents

1 **INTRODUCTORY IDEAS** — 1
  1.1 Origins — 2
  1.2 Evolution of the atmosphere — 3
    1.2.1 The prebiotic atmosphere — 3
    1.2.2 Gaia — 5
  1.3 The present state of the atmosphere — 6
  1.4 The thermal reservoir — 8
    1.4.1 The barometric law — 8
    1.4.2 Local thermodynamic equilibrium — 12
    1.4.3 Limits — 13
  1.5 Solar photons — 15
    1.5.1 Photolysis — 15
    1.5.2 Chemical kinetics — 17
  1.6 Atmospheric motions — 20
  1.7 Reading — 20
  1.8 Problems — 21

2 **THE THERMAL RESERVOIR** — 23
  2.1 Heat, work, and state functions — 24
    2.1.1 Heat — 24
    2.1.2 Work — 24
    2.1.3 Energy and entropy — 26
    2.1.4 Hydrostatic constraints — 27
    2.1.5 Irreversible work — 28
  2.2 Entropy — 29
    2.2.1 The entropy function — 29
    2.2.2 Potential temperature — 30
    2.2.3 Conditions for adiabaticity — 32
    2.2.4 The adiabatic lapse rate — 33
    2.2.5 Static stability — 34
  2.3 Global balances — 36
    2.3.1 Observed balances — 36
    2.3.2 Inventories — 36
    2.3.3 Hydrostatic constraints — 39
    2.3.4 Kinetic energy — 40
    2.3.5 Total potential energy — 41
    2.3.6 Surface heat and entropy fluxes — 42
  2.4 The second law of thermodynamics — 44
    2.4.1 Energy and entropy — 44
    2.4.2 Entropy inventory — 45

		2.4.3	Carnot efficiency	45
		2.4.4	Available potential energy	46
		2.4.5	Availability	47
	2.5	Reading		48
	2.6	Problems		49
3	ABSORPTION AND SCATTERING			57
	3.1	Interaction coefficients and cross sections		58
	3.2	Absorption by gases		60
		3.2.1	Lines and continua	60
		3.2.2	Line shapes	65
		3.2.3	Line wings	68
	3.3	Extinction by particles		70
		3.3.1	Mie theory	71
		3.3.2	Rayleigh's theory for small particles	72
		3.3.3	Geometric optics and anomalous diffraction	76
		3.3.4	Large and small-particle extinction	82
	3.4	Reading		83
	3.5	Problems		84
4	RADIATIVE TRANSFER			89
	4.1	Fundamental ideas		90
		4.1.1	Radiance and the equation of transfer	90
		4.1.2	Radiation flux and heating rate	91
		4.1.3	The thermal source function	92
		4.1.4	The scattering source function	95
		4.1.5	The complete source function	97
		4.1.6	Nonequilibrium source functions	97
	4.2	Solutions to transfer problems		99
		4.2.1 The integral equation		99
		4.2.2 Doubling and adding		102
	4.3	Approximate methods		107
		4.3.1 The frequency integration		107
		4.3.2 Approximate differential equations		112
		4.3.3 Radiation to space		115
	4.4	Reading		116
	4.5	Problems		118
5	CONSTRAINTS ON THE THERMAL STRUCTURE			123
	5.1	Vertical structure		124
		5.1.1	Radiative equilibrium (approximate theory)	124
		5.1.2	Convection in the lower atmosphere	127
		5.1.3	A cumulus convection model	131

		5.1.4	The Chapman layer	135
	5.2	\multicolumn{2}{l}{Meridional structure}	137
		5.2.1	Emission temperatures	137
		5.2.2	Energy balance climate models	139
	5.3	\multicolumn{2}{l}{Climate sensitivities}	142
		5.3.1	A simple greenhouse model	142
		5.3.2	Human influence on climate	144
		5.3.3	Water-vapor feedback	145
		5.3.4	Clouds and climate	146
	5.4	\multicolumn{2}{l}{Remote sensing}	147
	5.5	\multicolumn{2}{l}{Reading}	150
	5.6	\multicolumn{2}{l}{Problems}	151

6 OZONE 159

- 6.1 A brief history 160
- 6.2 Photodissociation rates 163
- 6.3 Photochemistry of an oxygen atmosphere 166
  - 6.3.1 Chapman's theory 166
  - 6.3.2 Excited states of atomic and molecular oxygen 170
  - 6.3.3 Departures from equilibrium 171
- 6.4 Gas phase catalysis 173
- 6.5 Transport 177
- 6.6 Tropospheric ozone 181
- 6.7 Nonhomogeneous reactions 182
- 6.8 Reading 184
- 6.9 Problems 184

7 TOPICS IN TROPOSPHERIC CHEMISTRY 191

- 7.1 Oxidants 192
  - 7.1.1 The free radicals 192
  - 7.1.2 Smog-forming reactions 193
  - 7.1.3 Other reactions with oxidants 195
- 7.2 Sulfur 196
  - 7.2.1 Sources and sinks 196
  - 7.2.2 Reactions in the boundary layer 199
  - 7.2.3 Reactions in the troposphere and stratosphere 201
- 7.3 Carbon dioxide and the carbon cycle 204
  - 7.3.1 A perspective 204
  - 7.3.2 Reservoirs 206
  - 7.3.3 Cycles 209
  - 7.3.4 Global change 213
- 7.4 Reading 215
- 7.5 Problems 215

8 CLOUDS AND PRECIPITATION — 219

- 8.1 Introduction — 220
  - 8.1.1 Cloud forms — 220
  - 8.1.2 Microphysical properties of clouds — 221
  - 8.1.3 Condensation — 223
- 8.2 Equilibrium between a droplet and its vapor — 225
  - 8.2.1 Kelvin and Raoult relations — 225
  - 8.2.2 The Kohler relation — 227
- 8.3 Condensation and freezing nuclei — 228
  - 8.3.1 Size distribution of aerosols — 228
  - 8.3.2 Sources — 230
  - 8.3.3 Cloud condensation nuclei — 231
  - 8.3.4 Ice nuclei — 231
- 8.4 Droplet growth in water clouds — 232
  - 8.4.1 Condensation — 232
  - 8.4.2 Coalescence — 237
- 8.5 Optical properties — 240
- 8.6 Motions — 243
  - 8.6.1 Cumulonimbus — 243
  - 8.6.2 The cloud-topped boundary layer — 243
  - 8.6.3 Synoptic systems — 244
- 8.7 Precipitation — 245
  - 8.7.1 Rain, snow, hail — 245
  - 8.7.2 The Wegener-Bergeron process — 246
  - 8.7.3 Droplet coalescence — 248
- 8.8 Reading — 249
- 8.9 Problems — 249

9 THE PLANETARY BOUNDARY LAYER — 253

- 9.1 Concepts — 254
  - 9.1.1 Classification of regions — 254
  - 9.1.2 The logarithmic profile — 256
  - 9.1.3 Eddy fluxes — 256
  - 9.1.4 Boussinesq approximation — 258
  - 9.1.5 Eddy energy — 259
  - 9.1.6 The Ekman and the surface layer — 262
- 9.2 Closure — 263
  - 9.2.1 The moment equations — 263
  - 9.2.2 K-closure — 264
  - 9.2.3 Mixing above the surface layer — 265
- 9.3 Similarity — 266
  - 9.3.1 The Richardson number — 266
  - 9.3.2 Monin-Obukhov similarity — 268
- 9.4 The convective boundary layer — 269
  - 9.4.1 Structure — 269

CONTENTS   xi

        9.4.2   Assumptions   270

        9.4.3   The capping inversion   271

        9.4.4   Entrainment fluxes   272

   9.5   Reading   273

   9.6   Problems   274

SOLUTIONS TO PROBLEMS   277

A   NUMBERS AND UNITS   287

B   THERMAL STATE OF THE ATMOSPHERE   289

C   ATMOSPHERIC MOTIONS   293

D   SOLAR RADIATION   297

E   WATER   299

F   THE FLUID EQUATIONS   303

G   THERMODYNAMIC CONCEPTS   307

H   KINETIC RATES AND RATE COEFFICIENTS   313

I   CLOUD GENERA   315

J   PROPERTIES OF SPHERICAL DROPS   317

INDEX OF DEFINITIONS   319

CHAPTER 1

INTRODUCTORY IDEAS

In the longest view, the earth is passing through a series of disequilibrium conditions between the origin of the universe, 15 billion years (By) ago, and (in thermodynamic terms) a final state of maximum entropy, or heat death. Processes that have led to the present state of the atmosphere include the formation of a protoplanetary nebula, its subsequent cooling with condensation of small bodies, assembly of planets from these small bodies, and processes subsequent to planetary formation, including tectonics, escape of species back to space, and the development of life.

This brings us to the earth's atmosphere as it now is, and the task of explaining its physical, chemical, and dynamical properties. By far the most important factor that prevents the earth from reaching a final equilibrium state is the sun. Energetic photons from the sun can strongly excite atmospheric molecules. If this excitation is rapidly thermalized, its effect is that of a reversible heat source, and the atmosphere itself is in a state of local thermodynamic equilibrium, which allows the use of Boltzmann statistics, and permits certain thermodynamic and statistical approaches. However, even in the lower atmosphere, the validity of bulk equilibrium assumptions may be questioned under some circumstances.

Equilibrium assumptions are also invalid for minor, reactive chemical species, some of which, for example, ozone, have important ecological connections. Energetic solar photons can dissociate molecules into highly reactive components, and the chain of reactions leading back to equilibrium may involve many reactive species. This chain of reactions must be studied by the methods of chemical kinetics.

The sun is also fundamental to the existence of life on earth, and life, in its turn, has subtle but important effects on the atmosphere, for example, it is probably responsible for the existence of free oxygen.

1.1 Origins

Hydrogen and a small amount of helium were the first atoms in the universe. For 10 By, clouds of gas condensed to form stars; nuclear reactions in their interiors and in supernova explosions created the heavier elements, in an elemental mixture that is observed throughout the universe. Ninety-five percent of all stars, including the sun, have the composition shown in Table 1.1, the so-called *cosmic abundances*.

Table 1.1 Cosmic abundances

Element	Atomic number	Abundances
H	1	2.6×10^{10}
He	2	1.8×10^9
O	8	1.8×10^7
C	6	1.1×10^7
Ne	10	2.6×10^6
N	7	2.3×10^6
Mg†	**12**	1.1×10^6
Si	**14**	1.0×10^6
Fe	**26**	9.0×10^5
S	16	5.0×10^5
Ar	18	1.1×10^5
Al	**13**	8.5×10^4
Ca	**20**	6.3×10^4
Na	**11**	6.0×10^4
Ni	**28**	4.8×10^4
Cr	**24**	1.3×10^4
Mn	**25**	9.0×10^3
P	15	6.5×10^3

†**Bold face** indicates *rock-forming* elements; the remaining elements are *volatiles*. The abundances are molecular number densities based on an assumed value for **Si** $= 10^6$.

According to *Laplace's theory*, about 5 By ago the sun condensed out at the center of a thin, hot, spinning disc of interstellar material, out of which, in the course of time, the planets formed. The *protoplanetary nebula* was hot near to its center (about 600 K near to the earth, according to some investigators), but cool in the outer reaches of the solar system. The nebula slowly cooled. The history of its temperature, and the resulting composition of successive condensates, lie at the heart of our speculations

INTRODUCTORY IDEAS

about planetary formation.

The *rock-forming components* in Table 1.1 condensed out as small grains in the earliest stages. In the inner solar system (the outer planets probably behaved differently), these grains accreted rapidly into planetesimals, the planetesimals into asteroid-sized bodies (1 to 100 km in size), and the asteroids accreted very slowly into the planets. This process is still going on, as is demonstrated by the occasional meteorite that strikes the earth. The lunar surface preserves the record of intense epochs of meteoritic bombardment during the first 1.5 By of the existence of the solar system, at the end of which time the cool protoplanetary nebula had been swept away. Now, after 4.6 By, the process of planet formation is almost complete.

The atmosphere and the oceans in their present form are the results of more recent processes. Noncondensable gases, such as oxygen and nitrogen, could not be collected by asteroids and small bodies because they have insufficient gravitational attraction. Later, when the planets formed and the gravitational forces were strong enough to retain an atmosphere, the protoplanetary nebula had dispersed. However, the *volatile components* in Table 1.1 all form compounds that can condense and are incorporated into asteroids together with the rock-forming elements, to be released at a later time. This view is consistent with the existence of the *carbonaceous chondrites*, meteorites that are believed to have condensed early in the formation of the solar system; these meteorites contain rock-forming and volatile components in similar proportions to the earth as a whole.

According to this view of the formation of our planet, the volatiles from which the atmosphere formed have been outgassed from the crust and the mantle in the course of tectonic processes that continue to this day in the form of volcanoes, fumeroles, steam wells, and geysers.

1.2 Evolution of the atmosphere

1.2.1 The prebiotic atmosphere

In view of the overwhelming proportion of hydrogen in the protoplanetary nebula, any early atmosphere might be expected to have been far more reducing than the present atmosphere: Specifically, CH_4 might be expected in place of CO_2, NH_3 in place of N_2O, and there should be no free oxygen. This may have been so at the very beginning, but this strongly reducing atmosphere is thought to have been replaced early on by a weakly reducing atmosphere, similar in composition to present-day effluent from tectonic processes, principally H_2O, CO_2, N_2, and CO, but lacking free oxygen.

How this change came about is a matter for debate. One factor may have been the formation of the highly reduced iron core of the planet: By separating from the mantle it could leave the surface rocks in a more oxidized state than before. Another possibility is the escape of gases from the atmosphere back into space, after the nebula dissipated. Escape processes usually favor loss of the lightest gas, hydrogen. Hydrogen can be obtained from the photolysis of H_2O, NH_3, or CH_4. If the hydrogen escapes, the atmosphere is left in a more oxidized state.

Several escape mechanisms have been identified, of which two operate under conditions of thermal equilibrium, see §1.4.2. The remainder are nonthermal and depend upon the energy of an absorbed solar photon. In either case, for a molecule to escape, its kinetic energy, $\frac{1}{2}mv^2$, where v is the velocity and m is the molecular mass, must exceed the gravitational potential energy, $mg(r)r$, where g is gravity and r the distance from the earth's center. It is also necessary for the molecule to be at an altitude where the atmosphere is so rarified that it can travel outward to space with low probability of collisions (this defines the *exosphere*). The energy condition requires

$$v > v_e = \sqrt{2gr}, \qquad (1.1)$$

where v_e is the *escape velocity*. At 500 km, the escape velocity is 10.77 km s^{-1}.

For a gas in thermal equilibrium, the most probable molecular velocity is

$$u = \sqrt{\frac{2\mathbf{k}T}{m}}. \qquad (1.2)$$

If $u > v_e$ for the major atmospheric species, the atmosphere can flow outward, as a fluid. This is happening at the present time with the solar wind from the sun, for example.

For present-day exospheric temperatures of $\simeq 1000$ K, $u \sim 4.1$ km s^{-1} for atomic hydrogen. Rapid escape cannot occur, but there will be a small number of molecules on the tail of the Maxwellian distribution for which velocities exceed the escape velocity. These rare molecules can escape to space in a loss process called *Jeans escape*. It is possible for significant quantities of hydrogen to have been lost by Jeans escape in the time since the planet formed.

Turning to nonthermal processes, large energies are associated with solar wind protons, and with far ultraviolet solar photons. For nonthermal escape the trick is to transform such energies into the kinetic energy of one or two atmospheric molecules without sharing it widely. A number of ways of achieving this have been identified, some with sufficient energy to allow oxygen and nitrogen atoms to escape.

For both thermal and nonthermal processes, an escaping molecule must first reach the base of the exosphere. This involves molecular diffusion through the lower and middle atmosphere. This can be a very slow process. Calculations suggest that diffusion may even provide the strongest limit to the escape of gases to space in the present epoch.

INTRODUCTORY IDEAS

Whatever may be the truth about these interesting questions, it is general practice to accept the existence of a weakly reducing, prebiotic atmosphere and to consider the influence upon it of the evolution of life.[1]

1.2.2 Gaia

Gaia, the earth goddess, was evoked by James Lovelock to symbolize his proposal that Life, in a global sense, has created the existing atmosphere from primordial constituents, to suit its own requirements. This proposal falls foul of the accepted tenets of neo-Darwinism, but has focused attention on what is undoubtedly correct, that the atmosphere is strongly conditioned by events taking place in the *biosphere*. The fundamental importance of this point will be clearer in Chapter 7, but it can be made briefly from a discussion of the origin of atmospheric oxygen.

The prebiotic atmosphere contained no free oxygen, but large amounts of the element were bound up in water and carbon dioxide, the former mainly in the oceans, and the latter mainly in carbonaceous rocks; on the time scale of billions of years, these volatile reservoirs can move to and from the atmosphere and cannot logically be distinguished from the atmosphere itself.[2] The 0.2 bar of oxygen that is now in the atmosphere represents a departure from this prebiotic atmosphere, minor in extent, but crucial in its implications. To explain the presence of oxygen we must also account for the absence of the free hydrogen or carbon that must have been simultaneously released if the oxygen came from the decomposition of water or carbon dioxide.

An important datum is the rate at which atmospheric oxygen is now disappearing from the atmosphere because this defines the size of the source necessary to maintain it. The fastest loss process that is known is the oxidation of surface rocks. The rate of this process depends on the rate at which weathering exposes reduced material in the rocks to the oxygen in the atmosphere. It is uncertainly estimated that all of the atmospheric oxygen could be consumed by this process in 4×10^6 y. This time sets limits to two processes. In the first place it must be possible to generate all of the atmospheric oxygen from water or carbon dioxide in this time or less; in the second place, the hydrogen or carbon released in the reaction must be removed from the atmosphere in this time or less.

Two possibilities exist for producing oxygen, photolysis and biospheric processes. There is no difficulty in producing oxygen in the required time by

---

[1] Exactly when life first appeared on earth is a matter for controversy. The earliest stromatalites are found in 3.5 By-old rocks, formed approximately 1 By after the earth first came together.

[2] If the water in the oceans and the carbon dioxide in carbonaceous rocks were to enter the atmosphere, it would contain 378 bar of water vapor and 38 bar of carbon dioxide.

the photolysis of carbon dioxide or water vapor, but the products containing excess hydrogen or carbon are gaseous, and the only proposal for removing any reduced gaseous constituent that has ever seemed remotely plausible is the removal of hydrogen to space (§1.2.1). Studies of the diffusion of hydrogen from the lower atmosphere to the exosphere all yield lifetimes for gaseous hydrogen of order of magnitude of the age of the planet, 10^3 times too slow to be consistent with the required oxygen source.

A biospheric source does not present these difficulties. Oxygen is released and carbon is fixed in green plants by the action of solar radiation. The entire oxygen reservoir can be supplied by this process in about 5×10^3 y, much faster than required by the weathering sink. Most of this oxygen will recombine with dead plant material, which is then recycled as CO_2, but reduced solid material, unlike gases, can be rapidly disposed of; it can be washed from the surface into the oceans where, at a sufficient depth, it will be isolated from the atmosphere for hundreds of millions of years. Some free oxygen will then be left in the atmosphere. An amount of organic carbon sufficient to account for the observed atmospheric oxygen exists in carbonaceous sediments on the ocean bottom.

1.3 The present state of the atmosphere

The present chemical composition of the earth's atmosphere is indicated by the lower-atmosphere *volume mixing ratios*[3] in Table 1.2, and the vertical distribution of these quantities from the surface to 120 km is shown in Figure 1.1. In addition to their natural variability, some of these species are impacted by human activity. These data differ greatly from the cosmic abundances because of events during and after the planet's formation. The mass of the atmosphere is 8.81×10^{-7} of the mass of the entire planet.

Table 1.2 and Figure 1.1 do not include water vapor. Some details for this important molecule may be found in Appendix E. The troposphere is strongly influenced by the predominantly liquid surface of the planet. The oceans together provide 2.38×10^{-4} of the total mass of the planet.

The thermal structure of the atmosphere (Figure 1.2, Appendix B) reflects important physical processes that will be discussed in later chapters, and is used as the basis for atmospheric nomenclature. The temperature decreases with height in the *troposphere*. This contrasts with the increase of temperature with height in the *stratosphere*. A temperature maximum separates the stratosphere and the *mesosphere*. Above the mesosphere lies the *thermosphere* or the *ionosphere*, in which temperature again increases with height.

---

[3]For an ideal gas, volume, molecular and molar mixing ratios or proportions are the same. The symbol used is μ. $\mu_i = \frac{n_i}{n}$ where n_i is the molecular number density of species i, and $n = \sum_i n_i$.

INTRODUCTORY IDEAS

Table 1.2 The most abundant components of dry air

Molecule	Volume mixing ratio in the troposphere	Comments
N_2	7.8084×10^{-1}	Photochemical dissociation high in the ionosphere; mixed at lower levels
O_2	2.0946×10^{-1}	Photochemical dissociation above 95 km; mixed at lower levels
A	9.34×10^{-3}	Mixed up to 110 km; diffusive separation above
CO_2	3.45×10^{-4}	Slightly variable; mixed up to 100 km; dissociated above
CH_4	1.6×10^{-6}	Mixed in the troposphere; dissociated in the mesosphere
H_2	$\sim 5 \times 10^{-7}$	Variable photochemical product; decreases slightly with height in the middle atmosphere
N_2O	3.5×10^{-7}	Slightly variable at the surface; dissociated in the stratosphere and mesosphere
CO	7×10^{-8}	Variable photochemical and combustion product
O_3	$\sim 10^{-8}$	Highly variable; photochemical origin
$CFCl_3$ and CF_2Cl_2	$1 - 2 \times 10^{-10}$	Industrial origin; mixed in the troposphere; dissociated in the stratosphere

The emphasis of this book is on the regions below 80 km. These are increasingly referred to as the *lower atmosphere* (troposphere) and the *middle atmosphere* (stratosphere plus mesosphere); above is the *upper atmosphere*. This is the most satisfactory nomenclature, and it will usually be employed in this book.

Perhaps the most obvious feature of the lower and middle atmospheres is that they are in motion, and that the motions are unsteady. Local gusts and lulls form the short time scales, whereas longer time-scale variability is associated with fronts and migrating storms. The motions that are suggested by a satellite view of cloud forms offer few simple features. Some details on mean atmospheric motions are given in Appendix C.

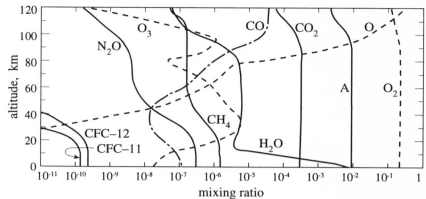

Figure 1.1 Vertical profiles of mixing ratios of selected species at an equinox. CFC-11 and CFC-12 are chlorofluorocarbons.

To complete this brief survey of the state of the atmosphere, we need to mention the important role of radiation. The source of heat and photochemical energy for the inner solar system is the sun (Appendix D). All other heat sources, such as internal heat from the planet, are small by comparison. The sun has an emission temperature of about 5783 K.

Radiation from the planet to space is either scattered solar radiation or *thermal radiation* (also called *planetary* or *terrestrial radiation*) emitted by the planet; planetary radiation has a mean emission temperature of about 250 K. Figure 1.3 shows black-body emission curves for 6000 K and 250 K together with some absorption properties of the atmosphere that help to define the interaction between the radiation streams and the atmospheric molecules.

1.4 The thermal reservoir

1.4.1 The barometric law

Classical approaches to physics and chemistry are usually in terms of *equilibrium systems*, systems that have had unlimited time in isolation to settle down. The laws of mechanics require that an equilibrium system have a unique pressure; the laws of thermodynamics require that it have a unique temperature, that it possess internal energy and entropy, and that the entropy is a maximum. Such ideas are fundamental to classical physics. However, real systems, such as the earth's atmosphere, do not exist indefinitely in isolation. Absorption of solar photons and atmospheric motions, amongst other phenomena, drive the atmosphere away from equilibrium.

INTRODUCTORY IDEAS

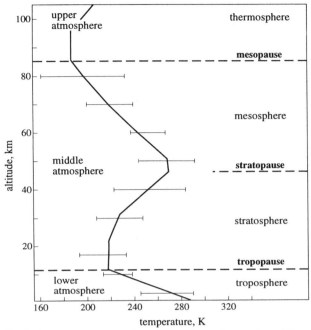

Figure 1.2 Vertical temperature structure of the atmosphere. The solid curve represents the U.S. Standard Atmosphere, see Appendix B. The horizontal bars show the range of monthly mean temperatures in the Northern Hemisphere.

But to provide a background to the discussion we first consider a simple model of the atmosphere in which the temperature and pressure are defined as a function of height (z) and motions are neglected.

The model atmosphere has a horizontal surface below and merges into space above; and we assume that variations of state parameters are much slower in the horizontal than in the vertical.[4] This is the *stratified atmosphere model* (see Figure 1.4). A gravitational force is necessary to constrain the atmosphere, and without solar radiation its temperature would rapidly fall to that of space.

In the absence of motions, the stratified atmosphere is in hydrostatic balance, that is, a balance between gravity and vertical pressure gradients (equation F.12). If we write $\rho = nm$ where n is the molecular number density and m is the molecular mass, equations (F.13) and (G.11) give,

$$\frac{d \ln p}{dz} = -\frac{1}{H}, \tag{1.3}$$

---

[4]It is tempting to say that the model is homogeneous horizontally, but that would be to ignore the slow horizontal variations of parameters that drive atmospheric motions.

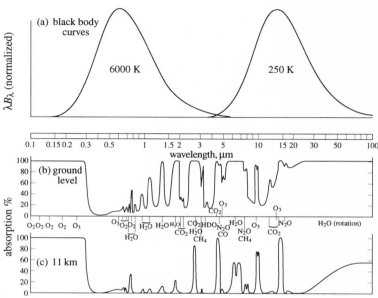

Figure 1.3 Atmospheric absorptions. (a) Black-body curves for 6000 K and 250 K. (b) Atmospheric absorption spectrum for a solar beam reaching the ground. (c) The same for a beam reaching the temperate tropopause. The areas beneath the curves in (a) are proportional to energy fluxes. Integrated over all angles, and averaged over time and over the globe, solar and terrestrial fluxes must balance; for this reason the two curves in (a) are drawn with equal areas. Conditions are appropriate to middle latitudes, with a solar elevation of about 40°, or for diffuse radiation.

where,

$$H = \frac{kT}{mg} = \frac{kT}{\mathcal{M} m_H g} \tag{1.4}$$

is the *atmospheric scale height*, $\mathcal{M}$ is the molar mass, m_H is the mass of a hydrogen atom, and **k** is Boltzmann's constant. Scale heights for some common gases at lower atmosphere temperatures are given in Table 1.3.

Lower and middle atmosphere temperatures do not depart by more than ±30% from 250 K, while pressures vary by a factor of 10^5 in the same regions of the atmosphere. As an approximation, therefore, we may integrate the hydrostatic relation at constant temperature,

$$p(z) \approx p_0 e^{-\frac{z}{H}}, \tag{1.5}$$

or

$$n(z) \approx n_0 e^{-\frac{z}{H}}, \tag{1.6}$$

where the zero suffix indicates the level $z = 0$.

INTRODUCTORY IDEAS

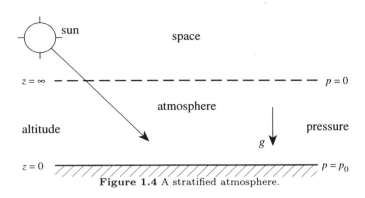

Figure 1.4 A stratified atmosphere.

Table 1.3 Atmospheric scale heights

Gas	Molar mass a.m.u.	Scale height, km 200 K	300 K
N_2	28.01	6.05	9.08
O_2	32.00	5.30	7.95
O	16.00	10.59	15.89
A	39.95	4.24	6.36
CO_2	44.01	3.85	5.78
H_2	2.02	83.91	125.87
H_2O	18.03	9.40	14.10
Air (dry)	28.96	5.85	8.78

Dalton's law requires that each gas in a gas mixture behave as if it alone occupied the entire volume. If left to come into complete equilibrium, each gas would have a scale height corresponding to its own molecular mass. From equation (1.6) and the data in Table 1.3, it follows that the molecular mixing ratio can change rapidly with height. However, for carbon dioxide, argon, and molecular oxygen (see Figure 1.1), this is not the behavior observed below 100 km: These gases have a common scale height corresponding to the mean molar mass for all atmospheric species (28.965 for dry air). The same is true below 15 km for nitrous oxide, methane, and the chlorofluorocarbons.

This behavior should not be a matter for surprise. If gases are to separate out from each other, it must be by molecular diffusion, a very slow process at normal pressures. In the lower atmosphere, atmospheric mixing overwhelms diffusion and establishes a constant mixing ratio for all gases. At the very low pressures prevailing in the upper atmosphere, diffusion is much more rapid and may dominate over mixing (see §1.4.3). If Figure 1.1 were slightly extended, it would show that above about 140 km species do, in the absence of competition from chemical reactions, follow their individual scale heights (*diffusive separation*), with the consequence that, well above 1000 km, hydrogen is the dominant atmospheric species, despite its small tropospheric mixing ratio ($\sim 5 \times 10^{-7}$).

In the lower and middle atmospheres, it is usual to look upon a uniform mixture of all gases except water vapor (which we refer to as *dry air*) as the equilibrium chemical state. Condensation of water and chemical reactions can then be regarded as perturbations on this equilibrium state.

1.4.2 Local thermodynamic equilibrium

An atmosphere in which temperature and pressure are functions of position is not in strict thermal equilibrium, but it is usually possible to divide the atmosphere into small subsystems each of which is effectively isothermal and isobaric, and may be treated as if in thermal and dynamical equilibrium. Each such subsystem is said to be in a state of *local thermodynamic equilibrium*.

The concept of local thermodynamic equilibrium is fundamental to atmospheric studies for the following reason. Each molecule of each chemical species can exist in a large number of energy states: translational, rotational, vibrational, and electronic. Each state is quantized and its properties may be stated in terms of a set of quantum numbers. If we know the populations that go with every possible set of quantum numbers, the state of the whole system is defined; but it would be impossible to work with the enormous number of variables involved. We often rely on equilibrium relations between state populations and on a statistical treatment of the system as a whole, using only a few macroscopic state variables, for example, pressure, temperature, and concentrations. The simplification is obvious, but under what conditions is it justified?

At the molecular level, the statistical relationship that allows us to derive macroscopic quantities is *Boltzmann's law*,

$$\frac{n_1}{n_2} = \frac{g_1}{g_2} \exp\left(-\frac{e_1 - e_2}{kT}\right), \qquad (1.7)$$

where the subscripts represent two states, n is the number density, g is the statistical weight, and e is the energy. If we wish to know whether macroscopic equilibrium relationships (e.g., the gas laws, the fluid equations, or Planck's radiation law) are valid, we must ask: What are the energy states that are involved, and do their populations follow Boltzmann's law to a sufficient degree of approximation? The answer may differ for different

INTRODUCTORY IDEAS

macroscopic variables at the same level in the atmosphere, and for the same variable at different levels.

Boltzmann's distribution is the result of randomization (or *thermalization*) by molecular collisions. One consideration that affects the rate of thermalization is the energy available in a collision compared to the energy required to cause a transition between states. The mean translational energy of a molecule is $\frac{3}{2}kT$, equal to 6.2×10^{-21} J molecule^{-1} at 300 K. By definition of a collision, each collision will change translational modes and bring them toward thermalization. On the other hand, the lowest distinct electronic level of molecular oxygen, to take a specific example, is the $^3\Sigma_u^+$ level, with an energy of 8.2×10^{-19} J molecule^{-1}. The fraction of molecules with translational energies more than 130 times their mean is extremely small, and it is a very rare collision indeed that is energetic enough to affect the population of this electronic level. Vibrational and rotational states lie in between translational and electronic states in this respect, with rotation usually requiring less energy per transition than vibration.

The possibility exists for thermalization to proceed rapidly for translational states while, at the same time, electronic states may be far from thermodynamic equilibrium. When this happens we may wish to retain the concept of "temperature" for translational levels (i.e., the quantity appearing in the Boltzmann relation), even though it may have no meaning for the electronic levels. A distinction may be drawn by use of the term *kinetic temperature*. Most references to atmospheric temperature are references to the kinetic temperature.

1.4.3 Limits

Thermal equilibrium among translational modes (and by implication, a kinetic temperature) may be shown to exist for essentially all conditions of importance to atmospheric physics and chemistry. We may demonstrate this proposition with two order-of-magnitude arguments.

First, molecules transfer properties from their origin over a distance equal to the molecular mean free path, λ. If $\lambda \ll H$ molecules perform collisions in what may be assumed to be an isobaric and isothermal environment, and we may expect approximate thermal equilibrium among the translational states; on the other hand, $\lambda \gg H$ is an almost certain recipe for a non-Maxwellian (i.e., non-Boltzmann) distribution. Figure 1.5 shows that $\lambda \approx H$ at 500 km. Despite uncertainties in this argument, this figure has significance because λ varies rapidly with height.

A similar conclusion may be reached by considering the disequilibrating effect of thermal diffusion. This may be confusing at first sight because thermal diffusion is usually regarded as an equilibrating process. However, in the context of local thermodynamic equilibrium, temperatures of

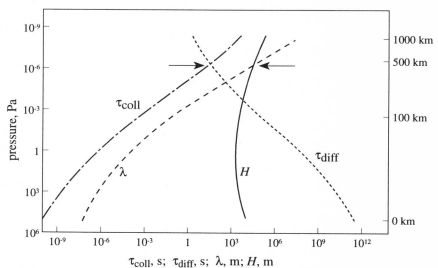

Figure 1.5 Scale height, molecular mean-free path, collision time, and diffusion time. The calculations are based on a collision diameter of 3.65×10^{-10} m and data from Table B.1.

contiguous subsystems can differ, and diffusion will mix molecules with differing translational statistics between them. Local collisions will then attempt to restore the local thermodynamic equilibrium. To order of magnitude, we may decide whether or not local equilibrium is preserved by comparing the characteristic time for heat diffusion (τ_{diff}) with the collision time (τ_{coll}): If the collision time is shorter, local equilibrium should prevail, and vice versa.

The characteristic rate of a thermal diffusion process over a vertical distance of a scale height is

$$\tau_{\text{diff}}^{-1} = \frac{K}{H^2} , \qquad (1.8)$$

where K is the thermometric diffusivity of air[5]. The two time constants are compared in Figure 1.5. We expect thermal equilibrium among the translational modes if $\tau_{\text{coll}} \ll \tau_{\text{diff}}$. The two time constants are approximately equal at 500 km. Consequently, at all lower levels translational modes of motion are in local thermodynamic equilibrium, and there is a well-defined, and measurable, kinetic temperature. This condition ensures the relevance of most familiar continuum concepts, for example, the Navier-Stokes equations for fluid flow.

---

[5]For a derivation of (1.8) see equation (9.37).

INTRODUCTORY IDEAS

All modes that are in thermal equilibrium communicate energy to each other through the translational modes. Together they form a *thermal reservoir*, whose properties are governed by the kinetic temperature. This is important for radiative source functions, which depend upon the state populations of nontranslational modes (see §4.1.5). If these nontranslational modes are not in equilibrium with the thermal reservoir, radiative transfer calculations can be very complex.

1.5 Solar photons

The sun is the strongest disequilibrating factor for the atmosphere. Whether and in what way solar energy enters the thermal reservoir, causes photochemical change, or affects processes through the agency of life, are questions of fundamental importance. Where photochemical changes are concerned, the laws of chemical kinetics must be used to follow the reactions.

1.5.1 Photolysis

For simplicity we consider the interaction between a solar photon, γ (energy, $h\nu$, where h is Planck's constant and ν is the frequency), and a diatomic molecule (AB). Bear in mind during this discussion that the solar photons in question are usually highly energetic, with wavelengths below 200 nm, in the ultraviolet spectrum. Thermal photons have wavelengths closer to 10 μm, and have energies at least 50 times smaller than those of the energetic solar photons. Chemical species that are stable in the laboratory in the presence of low-energy thermal photons may be disrupted by solar photons, leading to new species, that may be highly reactive and short-lived. We must trace the path of these disequilibrium species as they attempt to revert to the original state of *thermochemical equilibrium*.

An interaction with a solar photon may be represented by

$$AB + \gamma \rightarrow AB^* , \qquad (1.9)$$

where the asterisk represents a molecule in an excited state. If nothing else were to happen, the excited molecule would, after a sufficient number of collisions, end up in it original, unexcited state. There are many paths along which this de-excitation can take place. Table 1.4 shows four possible *initial steps*. There are, of course, many other possibilities, but these four illustrate the situation. The right arrows indicate that all of these processes are far from equilibrium, and take place in one direction only.[6]

---

[6] If the reaction takes place in both directions, as will happen at equilibrium, we use the symbol, $\rightleftharpoons$.

Table 1.4 Processes following an interaction $AB + \gamma \to AB^*$

1.	$AB^* \to AB + \gamma$	*radiative decay*
2.	$AB^* + M \to AB + M + e^\dagger$	*thermalization* or *quenching*
3.	$AB^* \to A + B$	*dissociation*
4.	$AB^* + C \to A + BC$	*reaction*

$^\dagger e$ represents excess energy.

Process 1 results in the re-emission of the original photon, but in a different direction and with a different state of polarization. This is *radiative decay*, also called *scattering* in radiative transfer theory (see §3.1). It is also possible for the emitted photon to be of lower energy than the incident photon, in which case AB will be left in a lower state of excitation.

Process 2 is *thermalization* or *quenching*. M represents any molecule that can carry away energy (e) and momentum during a collision. Processes 1 and 2 both end up with the molecule in its original state, the difference being that, for Process 1, the excess energy is carried away by the photon whereas, for Process 2, the energy is transferred to the thermal reservoir. The difference is crucial for studies of the thermal state of the atmosphere.

Processes 3 and 4 are *photochemical processes* (*photolysis*). If A, B, and BC are new species, not observed in thermochemical equilibrium, it follows that an energy barrier or *activation energy* must be involved that is larger than the energy of thermal photons but smaller than that of solar photons. For molecular oxygen, Process 3 requires a photon of frequency[7] greater than 41,660 cm^{-1}. For ozone the energy threshold is much lower, corresponding to 8,656 cm^{-1}.

The new species resulting from Processes 3 and 4 are often highly reactive, and they may be in excited energy states, with greater than normal reactivity. Processes 3 and 4 are usually only the first steps in a series of reactions. If they continue for long enough, the energy of the original photon will eventually be widely dispersed, and thermalized. Thus, from a thermal standpoint, Processes 2, 3, and 4 all have the final result that the solar energy ends up in the thermal reservoir; from the point of view of radiative transfer theory all three are called *absorption* processes.

Whatever the process, the time to achieve thermalization (the *thermalization time*) will be proportional to the time between collisions, that is,

---

[7] The energy of a photon is proportional to its frequency, and frequency is, therefore, the unit of choice when discussing photolysis. In practice, instead of frequency, spectroscopists prefer wavenumber, measured in cm^{-1}, and equal to the inverse of wavelength expressed in centimeters. 41,660 cm^{-1} corresponds to 242.4 nm, in the near ultraviolet spectrum; 8,656 cm^{-1} corresponds to 1155.3 nm, in the near infrared spectrum.

inversely proportional to the pressure; thermalization is more rapidly accomplished at high than at low pressures. If our concern is whether the solar energy ends up in the thermal reservoir, the competition is between Process 1 on the one hand and Processes 2, 3, and 4, on the other. There is a *natural decay time* associated with Process 1, which is a function only of the physical properties of the molecule, and does not depend on the pressure.

Thus, the ultimate fate of the solar photon will depend on the nature of the molecule and the pressure. If the pressure is high enough, the photon is thermalized, while if the pressure is low enough the photon is scattered. In radiative transfer theory absorption and scattering are treated as if they were intrinsic properties of the interaction between radiation on matter, but we now see that this is an oversimplified view and that the physical state of the atmosphere must be known before this judgement can be made.

To illustrate this point, consider two glass flasks containing gaseous chlorine and sodium vapor, respectively, the temperatures being adjusted until the molecular number densities are approximately equal. Both gases absorb yellow light. Illuminate both with intense beams of light from the sodium D lines. Their behavior is entirely different: The sodium flask glows uniformly from the light of multiply scattered photons; the chlorine flask emits no visible light—the photons are absorbed and their energy enters the thermal reservoir. This difference is determined by the natural decay times of the excited states (Process 1, Table 1.5), the sodium decay being by far the faster.

To return to the products of photolysis (Processes 3 and 4), the reactive species may have to make many collisions with "air" molecules before they find a reaction partner. The hydroxy radical (OH) has one of the shortest lifetimes that we shall discuss, about 0.25 s, in which time it makes $\sim 10^{10}$ collisions with other molecules. Any disturbance to the kinetic energy of OH will be thermalized long before a reaction can take place. Thus, reactions normally take place between species that are in thermal equilibrium even when they are not in chemical equilibrium. Because of this, reaction rate coefficients depend only upon the temperature of the environment, and not on the detailed motions of the molecules—a fortunate circumstance for the atmospheric chemist.

1.5.2 Chemical kinetics

The kinetic relationships that govern chemical reactions are reviewed in Appendix H. Most important atmospheric reactions involve only simple molecules. Nevertheless, there may be many unsuspected pathways between the initial reactions of Table 1.4 and the final products. To illustrate this point, consider the high-temperature reaction between methane and

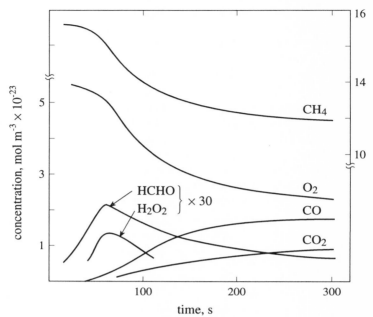

Figure 1.6 Oxidation of methane at 500°C.

oxygen, leading finally to the products carbon dioxide and water,[8]

$$CH_4 + 2O_2 \to CO_2 + 2H_2O \,. \tag{1.10}$$

If equation (1.10) were the only pathway from the species on the left to the species on the right, the reaction would be of third order, with the rate of loss of methane proportional to the methane number density and to the square of the oxygen number density. The rate of loss of methane should, therefore, *decrease* uniformly with time as methane and oxygen are used up. This is not what happens.

Figure 1.6 shows the result of mixing methane and oxygen in the laboratory at a temperature of 500°C. We immediately note two important features. First, the rate of disappearance of methane and oxygen *increases* with time between 20 and 70 s. Second, traces of hydrogen peroxide (H_2O_2) and formaldehyde (H_2CO) make transient appearances. It appears that (1.10) is kinetically irrelevant. The course of the reaction is controlled by intermediary steps.

---

[8]Methane oxidation in the atmosphere is important, but it takes place at a lower temperature, and with different reaction paths. This high-temperature example has been chosen for pedagogical purposes, because it is well documented.

INTRODUCTORY IDEAS

Table 1.5 Reaction mechanisms for the thermal oxidation of methane

(a)	$CH_4 + O_2 \rightarrow HO_2 + CH_3$	Initiation step
(b)	$CH_3 + O_2 \rightarrow CH_2O + OH$	} First chain
(c)	$OH + CH_4 \rightarrow H_2O + CH_3$	
(d)	$OH + CH_2O \rightarrow H_2O + CHO$	} Second chain
(e)	$CHO + O_2 \rightarrow HO_2 + CO$	
(f)	$HO_2 + CH_2O \rightarrow H_2O_2 + CHO$	
(g)	$H_2O_2 + M \rightarrow 2OH + M$	Chain branching
(h)	$OH + CO \rightarrow CO_2 + H$	} Third chain
(i)	$H + O_2 + M \rightarrow HO_2 + M$	
(j)	$HO_2 + CO \rightarrow CO_2 + OH$	
(k)	$OH + HO_2 \rightarrow H_2O + O_2$	} Chain termination
(l)	$2HO_2 \rightarrow H_2O_2 + O_2$	
(m)	Loss of HO_2 and H_2O_2 at the walls of the vessel	

Table 1.5 shows the minimum number of reactions needed to explain these laboratory data. Many combinations of carbon, hydrogen, and oxygen make their appearance. Hydrogen peroxide and formaldehyde are relatively stable species, and because of this they can build up measurable concentrations. Other species—CH_3, CHO, OH, and HO_2—are highly reactive. These are the *free radicals*. They have short lifetimes, exist in very low concentrations, but are essential to an understanding of the process. As long as free radicals are present, reactions will continue.

When free radicals react, they often produce new free radicals and, after a chain of reactions, the original free radical may reappear. When this happens, the free radical is acting as a *catalyst*, a species that participates in a reaction, but is not consumed by it. Processes (b)+(c), (e)+(f) and (h)+(i)+(j) are examples of chain reactions with the free radicals CH_3, CHO, and OH as catalysts. Reactions (k) and (l) remove OH and HO_2 radicals, yielding stable species H_2O and H_2O_2. These two reactions are *chain-terminating steps* that, directly or indirectly, stop all of the chain reactions.

We shall not discuss chemical kinetics further at this point: Issues will emerge when we discuss individual reactions in Chapters 6 and 7. In the final analysis, atmospheric chemistry is a very practical topic, depending upon the identification of reaction paths, and measurements in the laboratory of their rate coefficients.

1.6 Atmospheric motions

The final and perhaps least dramatic step in the thermalization of solar radiation is the dissipation of atmospheric motions, a process that also feeds small amounts of energy into the thermal reservoir. After local thermodynamic equilibrium has been established, and after the death and decay of living organisms, there will still remain a small degree of organization of heat sources in the atmosphere. More solar energy is deposited at low than at high latitudes, and more solar energy is deposited at the ground than in the atmosphere itself. This organization of heat sources leads to pressure gradients, and hence to motions.

Motions transport energy and species. Vertical motions, in particular, lead to condensation of water and ice and, sometimes, to precipitation. These questions belong to dynamical meteorology. To judge by societal investment, they may be the most important questions that atmospheric science has to deal with.

While dynamical meteorology per se is not treated in this book, some aspects of motions and transport may be treated in terms of local, steady-state models, and are included. Among these are simple climate models (Chapter 5), clouds and precipitation (Chapter 8), and interactions between the surface and the lower atmosphere (Chapter 9).

1.7 Reading

The evolution of the atmosphere is discussed by

Walker, J.C.G., 1977, *Evolution of the atmosphere*. New York: McMillan. More recent views are contained in articles by D.M. Hunten, J.F. Kasting, D. Ragnaud et al., and E.T. Sundquist in a 1993 issue of *Science* **259**, 915–941.

An overview of atmospheric physics and chemistry with a planetary perspective is given by

Chamberlain, J.W., and Hunten, D.M., 1987, *Theory of planetary atmospheres: An introduction to their physics and chemistry*, 2nd ed., New York: Academic Press.

Chemical processes are described by

Warneck, P., 1988, *Chemistry of the natural atmosphere*. New York: Academic Press, and

Brasseur, G., and Solomon, S., 1984, *Aeronomy of the middle atmosphere*. Dordrecht: D. Reidel.

INTRODUCTORY IDEAS

An early book on Gaia is

Lovelock, J.E., 1979, *A new look at life on earth*. New York: Oxford University Press.

Other topics are referenced in more detail in later chapters.

1.8 Problems

1.1 Barometric law. In the middle and lower atmospheres it is rarely essential to treat gravity as a variable. However, gravity varies as the inverse square of the distance from the center of the earth and, in the outer atmosphere, this is important:

$$g(z) = g_0 \frac{R_e^2}{(R_e + z)^2},$$

where g_0 is the surface gravity and R_e is the radius of the planet.

(i) Integrate equation (1.3) for a constant scale height and calculate the asymptotic pressure for $T = 600$ K, $p_0 = 10^5$ Pa, and for (a) $\mathcal{M} = 1$, (b) $\mathcal{M} = 29$. What conclusion do these figures suggest about the importance of hydrogen in the upper atmosphere?

(ii) At 1000 km, $\mathcal{M} \approx 1$ and $p \approx 10^{-8}$ Pa. Calculate the asymptotic pressure for $T = 600$ K.

1.2 The critical level. The *critical level* (the *escape level* or the *exobase*) refers to the level from which an upward-traveling molecule can escape to space with a low probability of collisions. We may define the level by requiring that there be one collision, on average, so that the probability of escaping to space is e^{-1}.

(i) Obtain an expression for the pressure at the critical level, p_e, in terms of the collision diameter, σ, the molecular mass, m, and gravity, g.

(ii) Show that the expression in (i) is equivalent to $\lambda = H$, where λ is the mean free path for a horizontal path, and H is the scale height.

(iii) Calculate p_e for $\sigma = 3.65 \times 10^{-10}$ m, $m = 29\mathbf{u}$ kg, and $g = 9.81$ m s^{-2}. **u** is the atomic mass unit (essentially the mass of a hydrogen atom). Interpolate the height of the critical level from the data in Table B.1.

1.3 Jeans escape. Jeans' theory leads to the following expression for the escape flux of molecules from the critical level,

$$F_e = \frac{n_e u}{2\pi^{1/2}} e^{-v_c^2/u^2} \left(\frac{v_e^2}{u^2} + 1 \right),$$

where n_e is the number density of molecules at the escape level, and v_e and u are defined by equations (1.1) and (1.2). There are difficult problems concerning the relationship of n_e to the total atmospheric inventory of the species concerned, but we may evaluate the escape mechanism itself by assuming that the lifetime of a species varies with $\frac{n_e}{F_e}$. Use this assumption to calculate, to order of magnitude, the ratio of lifetimes of atomic oxygen and atomic hydrogen at an exospheric temperature of 1000 K.

1.4 **Local thermodynamic equilibrium.** We make the assumption that energy is the only factor determining rearrangement of rotational levels. This is not strictly correct because other efficiency factors are also involved. However, with this assumption we may decide whether rotational energies are in local thermodynamic equilibrium by calculating the relative sizes of kinetic and rotational quanta of energy. (i) What is the most probable molecular translational energy at 1000 K? (ii) What is the energy of a typical rotational quantum of frequency 200 cm^{-1}? (iii) On the basis of these data, to what height in the atmosphere do you expect to find local thermodynamic equilibrium amongst rotational levels?

1.5 **Photodissociation.** Calculate the wavelength in nanometers below which photodissociation can occur for the reactions listed in the table. It does not follow that photodissociation *will* occur at this wavelength; it is also necessary for the photon to be absorbed. Figure 6.5 may be interpreted as giving the level above which photons of different wavelengths will be present in the atmosphere. From this figure, make an estimate of the levels at which photodissociation can occur in each case in the table.

Reaction	Dissociation energy, eV
$H_2 + \gamma \rightarrow H+H$	4.479
$N_2 + \gamma \rightarrow N+N$	9.762
$O_2 + \gamma \rightarrow O+O$	5.117
$O_3 + \gamma \rightarrow O_2+O$	1.052
$ClNO_3 + \gamma \rightarrow ClO+NO_2$	1.130
$CO_2 + \gamma \rightarrow CO+O$	5.455

1 eV = 1.602×10^{-19} J.

CHAPTER 2

THE THERMAL RESERVOIR

The meteorological literature is concerned in many different ways with relationships between motions and the state of the thermal reservoir. Figure 2.3 is an example. It represents, in schematic form, reservoirs of thermal and mechanical energies, integrated over the entire atmosphere, and the interactions between these reservoirs. It raises some questions.

Internal energy is changed by both heat and work interactions. There are several heat interactions in Figure 2.3, but why is there only one weak work interaction (the wind stress), and why should that manifest itself only through the kinetic energy? What exactly do the arrows mean? They look like fluid transfers, and some are labelled as "heat fluxes." But heat is a local interaction and not a state property that can move with a fluid, despite the historic caloric misrepresentation. And what is meant by "sensible heat"? Since latent heat is a form of internal energy, why does it have a separate box? One of the principal objectives of this chapter is to understand exactly what all of these features mean.

Entropy plays an unexpected role in atmospheric dynamics. The dynamical equations require, for closure, a single thermodynamic equation. Meteorologists use either an energy equation or an entropy equation, often the latter. What happened to the distinction between the first and second laws of thermodynamics if entropy and energy can be used interchangeably? Is the second law used to obtain the arrows between the "internal + potential energy" and the "kinetic energy" boxes in Figure 2.3? It should be because a question of efficiency is involved, but it turns out that climate discussion usually hinges on energy functions, such as the sum of potential and internal energy, which cannot convey information on efficiency.

Regardless of the status of the second law, the entropy function is a convenient function to use in meteorological calculations. If motions are fast enough as, for example, in a weather system, the flow will be approximately isentropic. For vertical flows this leads to a characteristic vertical temperature profile, called the adiabatic lapse rate, that offers a valuable first-order view of the tropospheric thermal structure. The adiabatic lapse rate also provides a criterion for the static stability of the atmosphere. Phase changes of water profoundly affect entropy in the lower atmosphere, and hence they also affect the adiabatic lapse rate and the static stability.

Chapter 2 explores these topics in a systematic development of the laws of thermodynamics as they apply to a fluid. The discussion is sometimes intricate, but that is the nature of the subject; it can be taken one section at a time. To follow this development it is essential for the reader to be confident with the fundamentals of thermodynamics and dynamics as outlined in Appendices F and G.

2.1 Heat, work, and state functions

2.1.1 Heat

The fluid equations differentiate between changes following the flow $(d\chi)$ and local changes $(\partial \chi)$, where χ is any specific[1] state function (see equation F.2). State functions are properties of the fluid and can be advected. Heat and work, on the other hand, are interactions, and their influence is local. We designate local rates of interaction (energy per unit time per unit mass) as $\dot{q}$ for heat and $\dot{\phi}$ for work; both have a number of components.

Two heat interactions are important in atmospheric physics: The first involves absorption and emission of radiation, and the second involves molecular diffusion or thermal conductivity. Later we shall find that molecular diffusion is only important at the earth's surface (§2.3.6) but, at this stage, we need not distinguish between these two interactions. We define an *interactive heat flux*, $\vec{F}_q$ from the relation,

$$\rho \dot{q} = -\nabla \cdot \vec{F}_q \,, \qquad (2.1)$$

where $\vec{F}_q$ is the sum of the radiative flux, $\vec{F}(\text{rad})$, and the diffusive flux, $\vec{F}(\text{diff})$, while ρ is the density. The definition of flux by equation (2.1) differs from the definition of a fluid flux, equation (F.21); the definitions correspond only if the velocities are all zero. It bears repeating that there is an important difference between state functions that can be transported by a fluid, and interactions that act locally.

2.1.2 Work

Consider a surface dA whose outward normal has direction cosines n_j. From the definition of the stress tensor, σ_{ij}, equation (F.5), the rate of working by the fluid on this surface is $u_i \sigma_{ij} n_j dA$, where u_i is a velocity

---
[1] The term *specific* means per unit mass, see Appendix G.

component. If we integrate over a closed surface enclosing a volume V we have, from Gauss' theorem, equation (F.19),[2]

$$\int_A u_i \sigma_{ij} n_j \, dA = \int_V \frac{\partial u_i \sigma_{ij}}{\partial x_j} \, dV \, . \tag{2.2}$$

Proceeding to the limit, $V \to 0$, gives the rate of working per unit volume,

$$\rho \dot{\phi} = \frac{\partial u_i \sigma_{ij}}{\partial x_j} = \sigma_{ij} \frac{\partial u_i}{\partial x_j} + u_i \frac{\partial \sigma_{ij}}{\partial x_j} \, . \tag{2.3}$$

The two terms on the right-hand side of equation (2.3) have important physical differences. The first term involves the velocity gradient but not the mean velocity of the fluid. Thus, it cannot change the velocity or the kinetic energy but because it distorts, compresses, or expands the fluid, it can change the internal energy, and it can do so either reversibly or irreversibly, because the stress tensor contains both a pressure and a viscous term, see equation (F.5). From the first term on the right of (F.5), and the first on the right of (2.3) we find the *reversible work*,

$$\rho \dot{\phi}_{\text{rev}} = -p \nabla \cdot \vec{u} = \frac{p}{\rho} \frac{d\rho}{dt} = -\rho p \frac{dv}{dt} \, . \tag{2.4}$$

The final term on the right of (2.4) is recognizable as a $-p\,dv$ work term, expressed per unit volume and per unit time. The equation of continuity, (F.1), was used to derive this equation.

From the second term on the right of (F.5), and with the help of the integral of (F.6), we find the *irreversible work* or *dissipation*,

$$\rho \dot{\phi}_{\text{irr}} = \tau_{ij} \frac{\partial u_i}{\partial x_j} \approx \rho \nu \left(\frac{\partial u_i}{\partial x_i} \right)^2 \geq 0 \, . \tag{2.5}$$

The dissipation term is positive definite, as we might expect.

The second term on the right of (2.3) depends on the velocity but not on its gradient; it translates but does not distort a small fluid element, and creates or destroys kinetic but not internal energy,

$$\rho \dot{\phi}_{\text{trans}} = u_i \frac{\partial \sigma_{ij}}{\partial x_j} \, . \tag{2.6}$$

Equation (2.6) can also be written as a sum of reversible and irreversible terms, but the distinction will not concern us here.

---

[2]See Lindzen (1990) for a more complete discussion. Following Lindzen we use either vector notation, as in equations (2.1) and (2.3), or indicial notation, as in equations (2.5) and (2.6), as convenient. The sum rule for repeated indices is used.

2.1.3 Energy and entropy

We are now in a position to derive expressions for the rates of change of the state functions. First, we introduce the *kinetic energy per unit mass*,

$$k = \frac{1}{2} u_i u_i = \frac{1}{2} \vec{u} \cdot \vec{u} \,. \tag{2.7}$$

Multiply the fluid momentum equation, (F.4), by u_i to give,

$$\rho \frac{dk}{dt} = -\rho g w + u_i \frac{\partial \sigma_{ij}}{\partial x_j} = -\rho g w + \dot{\phi}_{\text{trans}} \,. \tag{2.8}$$

This confirms our interpretation of equation (2.6) as concerning the kinetic energy only.

Potential energy is defined by,

$$\pi(z) = \int_0^z g(z')\,dz' \,, \tag{2.9}$$

where g is gravity. Hence,

$$\frac{d\pi}{dz} = g(z) \,. \tag{2.10}$$

Throughout this book we shall treat g as a constant, but it is a minor matter to avoid the approximation. One approach is to define *geopotential height*, $\tilde{z}$, such that,

$$g(z)\,dz = g_0\,d\tilde{z} \,, \tag{2.11}$$

where g_0 is standard gravity.

The vertical velocity is $w = \frac{dz}{dt}$, so that,

$$\rho \frac{d\pi}{dt} = \rho g \frac{dz}{dt} = \rho g w \,. \tag{2.12}$$

The rate of increase of kinetic energy, equation (2.8), can now be interpreted in terms of external work done on the fluid, equation (2.6), less increases in the potential energy, equation (2.12).

From the first law of thermodynamics, equation (G.1), and equation (2.4), the rate of change of *internal energy* is given by,

$$\rho \frac{de}{dt} = \rho \left(\dot{q} + \dot{\phi}_{\text{irr}} \right) + \rho \dot{\phi}_{\text{rev}} = \rho(\dot{q} + \dot{\phi}_{\text{irr}}) + \frac{dp}{dt} - \rho r \frac{dT}{dt} \,. \tag{2.13}$$

The ideal gas law has been used. r is the specific gas constant.

Enthalpy, the sum of internal energy and flow work, is widely used in fluid dynamics,

$$\rho \frac{dh}{dt} = \rho \frac{d(e + pv)}{dt} = \rho \frac{d(e + rT)}{dt} = \rho \left(\dot{q} + \dot{\phi}_{\text{irr}} \right) + \frac{dp}{dt} \,. \tag{2.14}$$

THE THERMAL RESERVOIR

Two summary statements may be obtained by adding energy equations

$$\rho \frac{d}{dt} \underbrace{(k + \pi + e)}_{total\ energy} = \rho \dot{q} + \rho \underbrace{(\dot{\phi}_{\text{trans}} + \dot{\phi}_{\text{rev}} + \dot{\phi}_{\text{irr}})}_{total\ work}, \qquad (2.15)$$

$$\rho \frac{d}{dt} \underbrace{(k + \pi + h)}_{energy+enthalpy} = \rho \dot{q} + \underbrace{\frac{\partial(\tau_{ij} u_i)}{\partial x_j}}_{viscous\ work} + \frac{\partial p}{\partial t}. \qquad (2.16)$$

Equation (2.15) is a statement of the first law of thermodynamics. The *viscous work* term in (2.16) is the sum of (2.5) and the viscous component of (2.6). For a steady state, ($\frac{\partial}{\partial t} = 0$), (2.16) is Bernoulli's equation: Mechanical energy plus enthalpy is conserved following the flow, unless changed by viscous work or external heating.

The rate of change of *entropy* is given by equations (G.5) and (2.14),

$$\rho \frac{ds}{dt} = \frac{\rho}{T} \left(\dot{q} + \dot{\phi}_{\text{irr}} \right). \qquad (2.17)$$

This is the simplest of the thermodynamic equations, particularly so if the term $\dot{\phi}_{\text{irr}}$ can be neglected (§2.1.5).

2.1.4 Hydrostatic constraints

Constraints are placed on the thermodynamic relations by the hydrostatic approximation. In its one-dimensional form, equation (F.13), the approximation requires,

$$d\pi = g\, dz = -\frac{dp}{\rho}. \qquad (2.18)$$

An important consequence of equation (2.18) is that pressure and height are no longer independent and that one degree of freedom has been lost from the system. Equation (2.18) allows potential energy to be combined with thermal energy in a simple way. Equation (2.13) may be written,

$$\rho \frac{d(e + \pi)}{dt} \approx \rho(\dot{q} + \dot{\phi}_{\text{irr}}) - \rho r \frac{dT}{dt}, \qquad (2.19)$$

and (2.14) may be written,

$$\rho \frac{d(h + \pi)}{dt} \approx \rho(\dot{q} + \dot{\phi}_{\text{irr}}). \qquad (2.20)$$

Two new state functions have been introduced that find frequent use in discussions of atmospheric thermodynamics; they are *total potential energy* $(e + \pi)$ and *static energy* $(h + \pi)$.

Equations (2.17) and (2.20) now give,

$$Tds \approx d(h + \pi) \,, \tag{2.21}$$

which implies that, under the hydrostatic approximation, the entropy provides no information that is not available from the static energy. This reflects a fundamental theorem in thermodynamics: The first and second laws are the same for a system with one or two degrees of freedom. The ideal gas laws have two degrees of freedom, the potential energy has one, and the hydrostatic relation removes one by relating height to pressure.

Another consequence of the hydrostatic approximation is to simplify the kinetic energy equation (2.8),

$$\begin{aligned}
\rho \frac{dk}{dt} &= -\rho g w - \vec{u} \cdot \nabla p + u_i \frac{\partial \tau_{ij}}{\partial x_j} \,, \\
&= -\vec{u} \cdot \nabla_h p - w \frac{\partial p}{\partial z} - \rho g w + u_i \frac{\partial \tau_{ij}}{\partial x_j} \,, \\
&= -\vec{u} \cdot \nabla_h p + u_i \frac{\partial \tau_{ij}}{\partial x_j} \,.
\end{aligned} \tag{2.22}$$

$\nabla_h = \frac{\partial}{\partial x} + \frac{\partial}{\partial y}$ is the horizontal part of the nabla operator. This approximation separates horizontal pressure forces, which can be directly measured, from vertical pressure forces, which cannot, and must be approximated. Large-scale motions are often approximated by the geostrophic equations, (F.14) and (F.15). Geostrophic flow is orthogonal to the pressure gradient so that for geostrophic flow the drive term (called the *ageostrophic drive*), $\vec{u} \cdot \nabla_h p$, is zero.

2.1.5 Irreversible work

Any one of the equations (2.13), (2.14), (2.17), (2.19), or (2.20) may be used as a thermodynamic heat equation to complement the equations of motion. Each involves the sum $(\dot{q} + \dot{\phi}_{irr})$. $\dot{q}$ may be calculated from the state of the atmosphere by the methods of Chapter 4. $\dot{\phi}_{irr}$ stands on the same footing as $\dot{q}$; Joule's famous experiments may be interpreted as establishing that their influence on the thermal state is identical. Although (2.5) is an explicit expression for $\rho \dot{\phi}_{irr}$, it is not always available because most of the contributions come from scales of motion less than 1 km in size that are neither routinely observed nor calculated theoretically in numerical models.[3] A common procedure is to neglect the irreversible work or, equivalently, to assume an inviscid atmosphere. Is this justified?

---

[3] A typical grid size for a numerical weather prediction model is 150 km; smaller scales must be approximately represented in terms of larger-scale parameters, an uncertain procedure, at best.

We may examine this question by comparing the column integrals of $\rho \dot{q}$ and $\rho \dot{\phi}_{\rm irr}$. The global average value of $\int_0^\infty \rho \dot{\phi}_{\rm irr}\, dz$ has been estimated from theory and observation, with great uncertainty, to lie between 1.5 and 3.5 W m^{-2}. A large contribution comes from the boundary layer, see §9.1.4. The heating term may be estimated from equation (2.1). Since fluxes are vertical in a stratified atmosphere, we may replace ∇ by $\frac{\partial}{\partial z}$, and

$$\int_0^\infty \rho \dot{q}\, dz = F_{q,z}(0) - F_{q,z}(\infty)\,. \tag{2.23}$$

The second term on the right side of equation (2.23) is numerically larger than the first. This term is displayed in panel (c) of Figure 2.2; its root-mean-square value, taken over the globe, is about 60 W m^{-2}. The neglected viscous term is, therefore, between 2.5% and 6% of another important term. These percentages will be larger in the boundary layer, where $\rho \dot{\phi}_{\rm irr}$ is concentrated.

2.2 Entropy

2.2.1 The entropy function

For most thermodynamic purposes water, in condensed or vapor phases, is the only important variable species in the atmosphere. Let m_v, m_c, and $m = m_v + m_c$ be the mass mixing ratios of vapor, condensate, and total water, respectively. In unit mass of the atmosphere, the mass of dry air (subscript a) is then $(1-m)$.

The entropy function may be written in two different forms, depending upon whether the water vapor is saturated or unsaturated; the two forms may be shown to be identical, but for an additive constant, if water vapor is treated as an ideal gas. For unsaturated air, assuming the mixture to behave as an ideal gas,

$$s = c_p \ln T - r \ln p\,. \tag{2.24}$$

c_p and r are the specific heat and specific gas constant for the gas mixture, see equation (G.18).[4] From (G.14),

$$r = r_a + m_v(r_v - r_a)\,, \tag{2.25}$$
$$c_p = c_{p,a} + m_v(c_{p,v} - c_{p,a})\,. \tag{2.26}$$

We indicate saturated air by using the *condensation temperature*, T_c, in place of the temperature. From equations (G.14), (G.18), (G.20), and

---
[4] Additive constants may be omitted. When expressions such as $\ln p$ and $\ln T$ are employed it is understood that the temperature and pressure have been divided by standard values.

(G.25),

$$s = (1-m)(c_{p,a}\ln T_c - r_a \ln p_a) + mc\ln T_c + \frac{m_v l}{T_c}, \qquad (2.27)$$

where l is the latent heat.

2.2.2 Potential temperature

The entropy function is commonly used to analyze atmospheric data, largely because of the simple form of equation (2.17), particularly when irreversible work is omitted, but it is usually replaced by a surrogate, the *potential temperature*, θ. The relationship between entropy and potential temperature is,

$$s = c_{p,a}\ln\theta. \qquad (2.28)$$

If we treat c_p as a constant,

$$\frac{ds}{dt} = \frac{c_{p,a}}{\theta}\frac{d\theta}{dt}. \qquad (2.29)$$

Adiabatic trajectories are, for most practical purposes, the same as isotherms of potential temperature.

Three approximations to (2.28) are in common use:
dry potential temperature,

$$\ln\theta_d = \ln T - \frac{r_a}{c_{p,a}}\ln p; \qquad (2.30)$$

virtual potential temperature,

$$\theta_v = \theta_d(1 + 0.61 m_v); \qquad (2.31)$$

equivalent potential temperature,

$$\theta_e = \theta_d \exp\left(\frac{l m_v}{c_{p,a} T_c}\right). \qquad (2.32)$$

Equations (2.30) and (2.31) apply to unsaturated air, while (2.32) is for saturated air. Equation (2.30) is an approximation for $m_v \ll 1$. (2.31) is approximately correct to first order in m_v. (2.32) can be used in saturated air; it neglects m_v, as does the dry potential temperature, but it includes the large term involving the latent heat term on the right side of (2.27) (magnitudes of terms are discussed at the end of §2.2.4). Throughout most of the literature the term *potential temperature* is taken to mean the dry potential temperature; the other approximations are given their full names.

But for an additive constant in the entropy, the dry potential temperature, θ_d, may be written in the form,

$$\theta_d = T \left(\frac{p}{p_0}\right)^{-r_a/c_{p,a}} . \qquad (2.33)$$

For dry air, $\frac{r_a}{c_{p,a}} = \frac{\gamma_a - 1}{\gamma_a} = 0.286$. According to (2.33) the dry potential temperature is the temperature that dry air would have if compressed adiabatically to pressure p_0. It is usual to choose $p_0 = 10^5$ Pa, and the potential temperature is then the temperature that a parcel of dry air will have if moved adiabatically to ground level.

Since the molar gas constant is universal, (2.25) can be written,

$$r = r_a \left\{ 1 + m_v \left(\frac{\mathcal{M}_a}{\mathcal{M}_v} - 1 \right) \right\} . \qquad (2.34)$$

$\mathcal{M}$ is a molar mass, and $\left(\frac{\mathcal{M}_a}{\mathcal{M}_v} - 1 \right) \approx 0.61$.

The molar specific heats of air and water vapor are similar but not identical. It is *approximately* true that,

$$c_p \approx c_{p,a}(1 + 0.61 m_v) . \qquad (2.35)$$

If we substitute (2.34) and (2.35) into the entropy expression (2.24), and treat $0.61 m_v$ as a small quantity,[5] we obtain (2.31) as an approximation.

This proliferation of potential temperatures harks back to the days before the existence of large computers, when atmospheric variables had to be easy to calculate. The dry potential temperature particularly commends itself since it is a simple function of temperature and pressure only. Figure 2.1 illustrates the use of dry potential temperature to describe the adiabatic trajectory of a parcel of air that rises from ground level (A), enters a cloud (at B), loses condensed water by precipitation (B to C), and then returns to ground level (C to D). On the legs AB and CD the dry potential temperature and the water vapor-mixing ratio are both constant; so too must be θ_e. Along the path BC the air is saturated, and θ_e is constant. It follows that the equivalent potential temperature is constant over the entire trajectory. From B to C the temperature decreases, which causes m_v to decrease rapidly, and the net effect is that $\frac{m_v}{T_e}$ decreases. It follows that $\theta_{d,2} > \theta_{d,1}$ and that, at the surface, $T_2 > T_1$. For the data given in the figure, the temperature increase is 15.8 K.

---

[5] Saturated air at a pressure of 10^5 Pa and a temperature of $30°C$ has a mixing ratio of 2.6%.

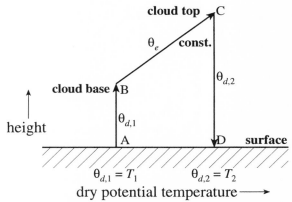

Figure 2.1 Schematic trajectory of an air parcel. If the conditions at B are: $z \approx 2$ km; $p = 7.950 \times 10^4$ Pa; $T = 275.15$ K; $m_v = 5.515 \times 10^{-3}$; then $\theta_{d,1} = 293.81$ K. If C is so high that essentially all the water is lost at this level, $\theta_{d,2} = 309.58$ K.

2.2.3 Conditions for adiabaticity

Adiabatic flows[6] are often invoked in heuristic discussions. In particular, a vertical adiabatic structure has some resemblance to the observed thermal structure, and may be used in simple atmospheric models. But first we may ask: Under what circumstances is the right side of equation (2.17) effectively zero? To answer this we compare the right side with the largest term on the left side. If our main concern is with vertical motions, adiabaticity may be assumed if,

$$\left| Tw \frac{\partial s}{\partial z} \right| = \left| \frac{c_p T w}{\theta} \frac{\partial \theta}{\partial z} \right| \gg |\dot{q}| , \qquad (2.36)$$

or, since $T \sim \theta$ in the troposphere,

$$\left| \frac{\dot{q}}{c_p} \right| \ll \left| w \frac{\partial \theta}{\partial z} \right| . \qquad (2.37)$$

A typical, average value of $\left| \frac{\dot{q}}{c_p} \right|$ is 1 K day^{-1} or 10^{-5} K s^{-1}, and a typical value of $\frac{\partial \theta}{\partial z}$ is 3.5 K km^{-1}. Equation (2.37) then requires,

$$|w| \gg 3 \times 10^{-3} \text{ m s}^{-1} .$$

This is a large value for the average vertical velocity on a global scale, but much larger vertical motions are observed on smaller scales such as

---

[6]Strictly speaking isentropic flows, but the distinction is academic if the dissipation rate is small.

THE THERMAL RESERVOIR

in weather systems, in convective clouds, and for small-scale mixing. For these smaller scales of motion we may expect to encounter approximate adiabatic conditions.

A similar analysis can be performed for horizontal motions, when the right side of (2.37) must be replaced by $\left|v\frac{\partial \theta}{\partial y}\right|$. If $\frac{\partial \theta}{\partial y} \sim 10^{-5}$ K m^{-1}, adiabaticity requires

$$|v| \gg 1 \text{ m s}^{-1}.$$

Once again, the inequality is obeyed for some motions, but not for others. Typical winds in a weather system, and the zonal winds away from the surface, are fast enough to be treated as adiabatic, but meridional motions in the atmospheric general circulation are too slow, see Appendix C.

2.2.4 The adiabatic lapse rate

For dry ascent $s(z)$ is a function of $p(z)$ and $T(z)$ only, and, from equation (2.30),

$$\frac{ds}{dz} = \frac{c_{p,a}}{\theta_d}\frac{d\theta_d}{dz} = \frac{c_{p,a}}{T}\frac{\partial T}{\partial z} - \frac{r_a}{p}\frac{\partial p}{\partial z}. \tag{2.38}$$

Vertical temperature gradients are generally negative, and it is common practice to introduce the *lapse rate*, $\Gamma = -\frac{\partial T}{\partial z}$. With this substitution, (2.38) becomes,

$$\frac{ds}{dz} = \frac{c_{p,a}}{T}(\Gamma_{\text{dry}} - \Gamma), \tag{2.39}$$

where

$$\Gamma_{\text{dry}} = -\frac{1}{\rho c_{p,a}}\frac{\partial p}{\partial z}, \tag{2.40}$$

is the *dry adiabatic lapse rate*. If, in addition, the atmosphere is in hydrostatic equilibrium, equation (2.18),

$$\Gamma_{\text{dry}} = \frac{g}{c_{p,a}} \approx 9.75 \text{ K km}^{-1}. \tag{2.41}$$

For dry adiabatic ascent, $\frac{ds}{dz} \approx 0$, and $\Gamma \approx \Gamma_{\text{dry}}$.

If the air is moist we write $T = T_c$, and from (2.32),

$$\begin{aligned}\frac{ds}{dz} &\approx \frac{c_{p,a}}{\theta_e}\frac{d\theta_e}{dz}, \\ &= \frac{c_{p,a}}{\theta_d}\frac{d\theta_d}{dz} + \frac{d}{dz}\left(\frac{lm_v}{T_c}\right), \\ &= \frac{\partial T_c}{\partial z}\left(\frac{c_{p,a}}{T_c} - \frac{l}{T_c}\frac{\partial m_v}{\partial T_c} - \frac{lm_v}{T_c^2}\right) - \frac{\partial p}{\partial z}\left(\frac{r_a}{p} - \frac{l}{T_c}\frac{\partial m_v}{\partial p}\right). \end{aligned} \tag{2.42}$$

Since,
$$m_v = \frac{\rho_v}{\rho} = \frac{p_v r}{p r_v} \approx \frac{p_v r_a}{p r_v}, \qquad (2.43)$$

we have,
$$\frac{\partial m_v}{\partial p} = -\frac{m_v}{r T_c \rho}, \qquad (2.44)$$
$$\frac{\partial m_v}{\partial T} = \frac{l m_v}{r_v T_c^2}. \qquad (2.45)$$

We have assumed $p \approx p_a$, and have used the approximate form, (G.26), of the Clausius-Clapeyron equation. We have also assumed that l is a constant.

We may combine (2.42), (2.44), and (2.45) with the hydrostatic approximation to give,
$$\frac{ds}{dz} = \frac{c_{p,a}}{\theta_e} \frac{d\theta_e}{dz} = \frac{\tilde{c}_p}{T_c} (\Gamma_{\text{sat}} - \Gamma), \qquad (2.46)$$

where
$$\tilde{c}_p = \left(c_{p,a} - \frac{l m_v}{T_c} + \frac{l^2 m_v}{r_v T_c^2} \right), \qquad (2.47)$$

and
$$\Gamma_{\text{sat}} = \frac{g}{\tilde{c}_p} \left(1 + \frac{l m_v}{r_a T_c} \right). \qquad (2.48)$$

If the atmosphere is saturated and motions are adiabatic,
$$\Gamma = \Gamma_{\text{sat}}. \qquad (2.49)$$

Γ_{sat} is the *saturated* or *moist adiabatic lapse rate*.

Although m_v is always small, quantities that multiply it in equations (2.47) and (2.48) are large. At 240 K, $\frac{l}{r_v T_c} \approx 26$, while $\frac{r_a}{c_{p,a}} \approx 0.3$. The third term in the parentheses in (2.47) is particularly large. Table 2.1 gives data on the relationship between Γ_{sat} and Γ_{dry}. In the upper troposphere there is little difference between them, but, near the surface in the tropics, the difference is large.

2.2.5 Static stability

We may expect to observe the adiabatic lapse rate in any well-mixed region of the atmosphere, an example of which is the diurnal, convective boundary layer, which we shall discuss in §9.4. Throughout most of the troposphere, however, the lapse rate is less than adiabatic, either dry or moist. Under this circumstance, the atmosphere is *stable* to small disturbances, as may be seen from the following argument.

Table 2.1 $\Gamma_{\text{sat}}/\Gamma_{\text{dry}}$ as a function of temperature and pressure

| Tempererature | Pressure, Pa | | |
K	10^5	5×10^4	2.5×10^4
300	0.392	—†	—
280	0.580	0.455	—
260	0.815	0.697	0.556
240‡	—	0.915	0.845
220	—	0.989	0.978

†Blanks indicate unlikely combinations of temperature and pressure.

‡Ice data were used for 240 K and 220 K; water data for higher temperatures.

Table 2.2 Conditions for stability

unstable	$\frac{d\theta}{dz} < 0,$	$\Gamma > \Gamma_{\text{ad}}^\dagger$
neutral	$\frac{d\theta}{dz} = 0,$	$\Gamma = \Gamma_{\text{ad}}$
stable	$\frac{d\theta}{dz} > 0,$	$\Gamma < \Gamma_{\text{ad}}$

†Γ_{ad} is either Γ_{sat} or Γ_{dry}, as appropriate.

In §9.1.4 we show that the condition for small disturbances to grow is that they carry an upward flux of entropy. Now suppose that $\frac{d\theta}{dz} < 0$. If this atmosphere is mixed strongly entropy will be carried upwards and, as a consequence, disturbances will grow. This condition is *unstable*. The converse is true for $\frac{d\theta}{dz} > 0$, which is *stable* to small disturbances. The adiabatic state itself is *neutral* (see Table 2.2).

The lapse rate may (and often does) lie between the dry and moist adiabatic lapse rates. In this event, a parcel of air in the dry region below a cloud would be stable to small disturbances. If it were forced upwards by external drives it would cool adiabatically and would eventually reach the condensation temperature; then it would enter the cloud and be unstable in the moist environment. The atmosphere is then said to be *conditionally unstable*.

2.3 Global balances

2.3.1 Observed balances

Figures 2.2, 2.3, and 2.4 show averaged data relevant to our discussion of global energy balances. The data are partly observed and partly inferred from other data. For example, fluxes of radiation to and from space can be measured directly from an earth satellite. On the other hand, the distribution with height of the radiative heating is theoretically derived from observed distributions of temperature and radiating gases (see Chapter 4 for methods of calculation).

Figure 2.2 gives information on radiative fluxes into and out of latitude zones and, from the balance, the average dynamical fluxes from one latitude to another. Figure 2.3 gives an overall summary of the fluxes of all forms of energy into and out of the entire atmosphere, together with some indication of fluxes between reservoirs of dry static energy, internal energy of condensation, and kinetic energy. Note that the energy transported to and from the kinetic energy reservoir is very small compared with other fluxes. Figure 2.4 shows the vertical distribution of the global average radiative heating and cooling rates, expressed as $\dot{q}/c_p$ in kelvin per day. $\dot{q}$ is the energy per unit mass and, because most of the atmospheric mass is in the lower atmosphere, so too is most of the absorption and emission of radiant energy. For global averaged data in the lower atmosphere, cooling by thermal radiation exceeds heating by solar radiation. The balance is provided by fluid transports that act in such a way as to carry energy from the surface, where solar radiation is mainly absorbed, up into the atmosphere.

2.3.2 Inventories

We shall employ brackets {} to indicate *inventories* or totals contained in a reservoir. The most interesting reservoirs are global, but the same definitions may be used for more limited systems, such as a vertical column of unit cross section. The definition is,

$$\{\chi\} = \int_{system} \rho \chi \, dV \, , \tag{2.50}$$

where χ is any specific function.

For a state function we have, from equations (2.50), (F.19) and (F.20),

$$\left\{\frac{d\chi}{dt}\right\} = \int_{surface} F_n(\chi) \, dA + \frac{\partial \{\chi\}}{\partial t} = \tilde{F}(\chi) + \frac{\partial \{\chi\}}{\partial t} \, . \tag{2.51}$$

$\tilde{F}(\chi)$ is the *outward* flux of χ integrated over the boundaries of the system. A global system has no lateral boundaries, and the relevant surfaces are

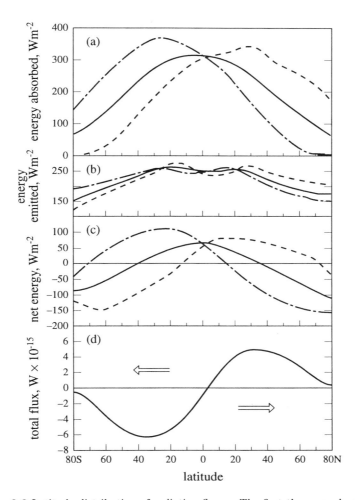

Figure 2.2 Latitude distribution of radiation fluxes. The first three panels give time- and-longitude averages of the radiation fluxes at the top of the atmosphere, as might be measured from a satellite. (a) is the net solar flux, incident less reflected. (b) is the outgoing flux of thermal radiation, and (c) is the balance, (a)−(b). (d) is the horizontal flux that must be carried across latitude circles by motions in the atmosphere-ocean system (northward positive) required to balance (c). Heavy, full lines are annual averages; dash-dot lines are averages over the three northern winter months; broken lines are averages over the three northern summer months.

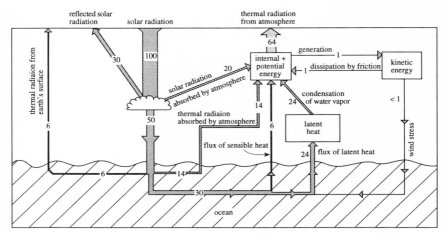

Figure 2.3 Energy fluxes. Fluxes are between the surface, the atmosphere, and space, and internally between major energy reservoirs. The data are averaged over the globe. Fluxes between boxes (reservoirs) are given as percentages of the incident solar radiation (100% = $\frac{1373}{4}$ W m^{-2}, see discussion in §5.1.1). The term "sensible heat" is used for the dry component of the internal energy, $c_v T$, and "latent heat" for the moist component, $m_v l$; see §2.3.6 for discussion.

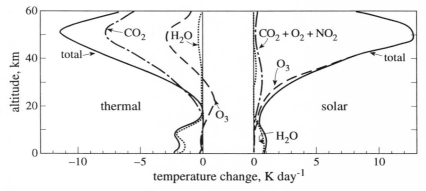

Figure 2.4 Distribution with altitude of globally averaged radiative heating rates, expressed as $\dot{q}/c_p$. The heating is broken down into solar and thermal radiation, and between species.

THE THERMAL RESERVOIR

at the earth's surface and in space; radiative energy is the only flux over the upper surface. Some investigators also omit lateral transports from systems that are less than global in extent, but that requires justification on a case-by-case basis.

This is an appropriate point to remind ourselves that the earth's fluid envelope consists of both the atmosphere and the oceans, and that they interact. The mass of the oceans is approximately 300 times greater than that of the atmosphere, and the rates of change in the oceans are slower than those in the atmosphere, depending upon the depth of the oceans involved. For example, the deep oceans circulate in about 1000 years, and it may take this long to distribute the results of a changed surface flux. We may write a balance equation similar to (2.51) for the oceans, with a surface flux term that is equal to, but of opposite sign than, that for the atmosphere. If the two equations are combined so that the entire fluid system is treated together, the surface flux terms vanish. Whether we treat the fluid earth as two systems or one is a choice for the investigator. For day-to-day problems of weather forecasting, the state of the oceans is usually regarded as given; for climate forecasting, particularly if large computers are available, the atmosphere and oceans may be treated as a single system.

2.3.3 Hydrostatic constraints

The hydrostatic relation imposes constraints on the inventory of the potential energy. If we adopt the simplest form of the approximation, equation (F.13), the inventory for an isolated vertical column over unit area is

$$\begin{aligned}
\{\pi\} &= \int_{atmos} \rho g z \, dV \,, \\
&= \int_0^\infty \rho g z \, dz \,, \\
&= \int_0^{p_0} z \, dp \,, \\
&= \int_0^\infty p \, dz + [(pz)_\infty - (pz)_0] \,, \\
&= r \int_0^\infty \rho T \, dz \,. \quad (2.52)
\end{aligned}$$

z_0 and p_∞ are both zero.

In the limit of m_v small ($m_v \ll \frac{c_p}{c}$), $c_{p,a}$ may be replaced by c_p, and equations (G.14), (G.16), and (G.21) give for the enthalpy inventory,

$$\{h\} = \{c_p T\} + \{m_v l\} \,, \quad (2.53)$$

where the first term on the right is the dry contribution to the enthalpy. The integration constant in (G.16) has been omitted and we have retained terms in m_v only when multiplied by l.

Similarly, from (G.14), (G.15), and (G.23),

$$\{e\} = \{c_v T\} + \{m_v l\} \,. \tag{2.54}$$

The dry contributions may be evaluated,

$$\{h_{\text{dry}}\} = \{c_p T\} = c_p \int_0^\infty \rho T \, dz \,, \tag{2.55}$$

$$\{e_{\text{dry}}\} = \{c_v T\} = c_v \int_0^\infty \rho T \, dz \,. \tag{2.56}$$

Hence, the potential energy, the dry enthalpy, and the dry internal energy inventories are not independent but have the ratios:

$$\{c_p T\} : \{c_v T\} : \{\pi\} = c_p : c_v : r \approx 7 : 5 : 2 \,. \tag{2.57}$$

A further result from the hydrostatic approximation is that, for either dry or moist conditions,

$$\{h\} = \{e + \pi\} \,. \tag{2.58}$$

The relative sizes of inventories that go with the boxes in Figure 2.3 have been calculated from observed data as percentages of the total energy, with the following results,

$$\{e + \pi + k\} = \{c_v T + lm + \pi + k\} \approx 1.3 \times 10^{24} \text{ J} \,,$$

$$\begin{aligned}
\{c_p T\} &= 97.5\%, \\
\{c_v T\} &= 70.4\%, \\
\{\pi\} &= 27.1\%, \\
\{lm_v\} &= 2.5\%, \\
\{k\} &= 0.05\% \,.
\end{aligned}$$

Note the very small size of the kinetic energy reservoir compared with the reservoirs of thermal and potential energy.

2.3.4 Kinetic energy

Equations (2.6), (2.8), and (2.51) give,

$$\frac{\partial \{k\}}{\partial t} = \left\{ \frac{u_i}{\rho} \frac{\partial \sigma_{ij}}{\partial x_j} \right\} - \{gw\} \,. \tag{2.59}$$

THE THERMAL RESERVOIR

Equation (2.59) does not contain a surface flux term for subtle reasons. The only boundary interactions on a closed thermodynamic system are heat and work. The former does not directly affect the kinetic energy. The latter involves surface friction that affects the momentum and, hence, indirectly affects the velocity and the kinetic energy. However, surface frictional forces are accounted for by the stress tensor, σ_{ij}, and frictional effects appear in equation (2.62). Surface friction is, therefore, accounted for.

From the definition of the stress tensor, equation (F.5),

$$\frac{\partial \{k\}}{\partial t} = -\left\{\frac{\vec{u}}{\rho} \cdot \nabla p\right\} + \left\{\frac{u_i}{\rho} \frac{\partial \tau_{ij}}{\partial x_j}\right\} - \{gw\} \ . \tag{2.60}$$

The second term on the right side of (2.60) may be written,

$$\left\{\frac{u_i}{\rho} \frac{\partial \tau_{ij}}{\partial x_j}\right\} = \left\{\frac{1}{\rho} \frac{\partial (u_i \tau_{ij})}{\partial x_j}\right\} - \{\dot{\phi}_{\text{irr}}\} = -\int_{surface} u_i \tau_{iz} dA - \{\dot{\phi}_{\text{irr}}\} \ . \tag{2.61}$$

The first term on the right side of (2.61) is frictional work performed at the surface, as discussed above (note the sign change because the direction of increasing z is the negative j direction). If the horizontal velocity components of the surface itself are nonzero, this term is finite. This applies to the oceans in regions of surface currents. This does not mean that surface friction over land has no influence on the kinetic energy, but only that it is contained in the term $\dot{\phi}_{\text{irr}}$.

From (2.60), (2.61), and the hydrostatic approximation (2.18), the rate of change of the kinetic energy inventory may be written,

$$\frac{\partial \{k\}}{\partial t} = -\int_{surface} u_i \tau_{iz} dA - \left\{\frac{\vec{u}}{\rho} \cdot \nabla_h p\right\} - \{\dot{\phi}_{\text{irr}}\} \ . \tag{2.62}$$

Other things being equal, the fluid momentum equation, (F.4), will lead to a negative correlation between velocities and pressure gradients. Consequently, the second term on the right-hand side of equation (2.62) is a generation term, while the third term is a sink. Each term on the right side of equation (2.62) is indicated by an arrow to or from the "kinetic energy" box in Figure 2.3: The first term is called "wind stress"; the second is called "generation"; and the third is called "dissipation by friction."

2.3.5 Total potential energy

We require the following two relationships. From the equation of continuity, (F.1),

$$p\rho \frac{dv}{dt} = -\frac{p}{\rho} \frac{d\rho}{dt} = p\nabla \cdot \vec{u} = \nabla \cdot p\vec{u} - \vec{u} \cdot \nabla p \ , \tag{2.63}$$

and from the definition of π, equation (2.9), and the hydrostatic approximation, equation (F.12),

$$\rho\frac{d\pi}{dt} = \rho g w = -w\frac{\partial p}{\partial z} . \qquad (2.64)$$

If we combine (2.13), (2.63), and (2.64), and introduce the horizontal divergence operator, ∇_h, we have

$$\rho\frac{d(e+\pi)}{dt} = \rho(\dot{q} + \dot{\phi}_{irr}) + \vec{u} \cdot \nabla_h p - \nabla \cdot p\vec{u} . \qquad (2.65)$$

The rate of change of the inventory is,

$$\frac{\partial\{e+\pi\}}{\partial t} = -\tilde{F}(e+\pi) + \{\dot{q}\} + \left\{\frac{\vec{u}}{\rho} \cdot \nabla_h p\right\} + \{\dot{\phi}_{irr}\} . \qquad (2.66)$$

For a global inventory the term $\int_{atmos} \nabla \cdot p\vec{u}\, dV = -\int_{surface} pw\, dA$ is zero because $w = 0$ at the surface, and $p = 0$ in space. From equations (2.23) and (2.50), $\{\dot{q}\}$ is the net heat flux from the global atmosphere.

Apart from the surface fluxes (which represent external interactions), the terms in the kinetic energy and total potential energy equations are the same but for a change of sign: A source for one is a sink for the other, and vice versa. In the absence of surface fluxes of total potential energy and of heat interactions, (2.62) and (2.66) lead to

$$\frac{\partial\{e+\pi\}}{\partial t} = -\frac{\partial\{k\}}{\partial t} . \qquad (2.67)$$

For a long-term average over the entire atmosphere, net surface fluxes must be zero if we are to maintain a steady state, and the relationship in equation (2.67) should hold.

2.3.6 Surface heat and entropy fluxes

The surface flux term, $\tilde{F}(e + \pi)$, in equation (2.66) is represented by two terms in Figure 2.3, the "flux of latent heat," and the "flux of sensible heat." Since the vertical velocity is zero at the lower boundary, neither of these fluxes is a fluid flux in the sense of equation (F.21). In a macroscopic sense, they are interactions by heat and matter external to the atmosphere. In a microscopic sense, they are the result of molecular diffusion through a very thin *molecular diffusion layer*, whose structure per se does not concern us.

From the additivity relation, (G.14), equations (G.16) and (G.23), and the definition of fluid flux, (F.21), we may write at the top of the molecular diffusion layer,

$$\begin{aligned}
\tilde{F}(e+\pi) &= \tilde{F}(e) , \\
&= \tilde{F}([1-m]e_{\text{dry}}) + \tilde{F}(m_v l) + \tilde{F}([m_v + m_c]cT) , \\
&\approx \tilde{F}(e_{\text{dry}}) + l\tilde{F}(m_v) + cT\tilde{F}(m) .
\end{aligned} \qquad (2.68)$$

m_v, m_c, and m are, respectively, the mixing ratios of the vapor, the condensed phase, and total water. Implicit in (2.68) is the assumption that all phases and components are in thermodynamic equilibrium, a questionable assumption for falling rain, but one that we shall adopt for simplicity. We have neglected $\tilde{F}(\pi)$ at the top of the diffusion layer; $F_z(\pi) = w\pi(\epsilon) = wg\epsilon \to 0$ as $\epsilon \to 0$.

The third term on the right of equation (2.68) contains the term $\tilde{F}(m)$, the net flux of water in all phases. For a steady state between the atmosphere and the surface, $\tilde{F}(m) = 0$. $-\tilde{F}(m_v)$ is the net flux of vapor from the surface to the atmosphere, or the *global evaporation rate*, and is, in the long run, equal but opposite in sign to the *global precipitation rate*. Since precipitation is part of the meteorological record, the second term on the right of (2.68) may be estimated (with some uncertainty) from that record. This term is the "flux of latent heat" in Figure 2.3.

The first term on the right of (2.68) is the "flux of sensible heat." It is not simply the heat conducted through the diffusion layer, as we may see by considering the vertical flux of dry enthalpy at the top of the diffusion layer, $z = \epsilon$,

$$F_z(h_{\text{dry}}, \epsilon) = \frac{c_p}{c_v} F_z(e_{\text{dry}}, \epsilon) . \qquad (2.69)$$

From equations (2.1) and (2.20) we may write, inside the diffusion layer, $0 < z \le \epsilon$,

$$\frac{\partial F_z(h_{\text{dry}})}{\partial z} \approx -\frac{\partial F(\text{diff})}{\partial z} . \qquad (2.70)$$

$F(\text{diff})$ is the diffusive heat flux, see the discussion of equation (2.1). Equation (2.70) is written for a steady state in the diffusive layer, ($\frac{\partial}{\partial t} = 0$), and only vertical flux components are considered. It also assumes that m_v does not diverge (no condensation), and that the divergence of the diffusive heat flux greatly exceeds both ϕ_{irr} and the divergence of the radiation flux. The large size of the diffusive flux divergence comes about because of the rapid changes of temperature gradients characteristic of a thin molecular boundary layer.

The picture presented by (2.70) is of a conductive heat flux leaving the surface with a value $F(\text{diff}, 0)$ and falling to zero at $z = \epsilon$, while the fluid flux of dry enthalpy starts from zero at the surface, and achieves a value of $F(\text{diff}, 0)$ at $z = \epsilon$, where it sets the boundary condition for the atmosphere above. With the help of equation (2.21), we have the following boundary conditions for fluxes at the top of the boundary layer,

$$F_z(h_{\text{dry}} + \pi, \epsilon) = F_z(h_{\text{dry}}, \epsilon) = F(\text{diff}, 0) , \qquad (2.71)$$

$$F_z(e_{\text{dry}} + \pi, \epsilon) = F_z(e_{\text{dry}}, \epsilon) = \frac{c_v}{c_p} F(\text{diff}, 0) , \qquad (2.72)$$

$$F_z(s_{\text{dry}}, \epsilon) = \frac{F(\text{diff}, 0)}{T_g} , \qquad (2.73)$$

where T_g is the temperature at the top of the molecular boundary layer. Surface fluxes are further discussed in Chapter 9.

We have now identified the only role of molecular heat and entropy diffusion in the lower atmosphere and, from here on, we shall, above the molecular diffusion layer, identify the nonfluid heat flux, $\vec{F}_q$, with the radiative flux, $\vec{F}(\text{rad})$.

All of the arrows and boxes in Figure 2.3 have now been identified and defined. The radiative fluxes are broken up into a number of components that will be discussed further in Chapters 3 and 4.

2.4 The second law of thermodynamics

2.4.1 Energy and entropy

The second law of thermodynamics does not occupy a prominent place in discussions of atmospheric processes. Although we introduced entropy in equation (2.17), and used it as a basis for defining potential temperature, all of our discussion in this chapter has been couched in terms of energy arguments, for which we only need to know the first law of thermodynamics. The reason for this, in terms of the limited number of degrees of freedom in the system, has already been discussed. If we wish to avoid the approximations involved, such as the hydrostatic approximation, the second law comes into its own.

The most important questions about climate that involve the second law are connected with the treatment of the dissipation term, $\dot{\phi}_{\text{irr}}$, which the kinetic equation, (2.62), shows to be the fundamental constraint on the growth of kinetic energy. On the basis of energy considerations alone, this quantity cannot be distinguished from heating, $\dot{q}$, since both quantities always occur together. The second law of thermodynamics, being concerned with efficiency of conversion of heat to work, must be introduced if the constraints on this term are to be understood in a way that is model-independent.

2.4.2 Entropy inventory

From equations (2.17), and (2.51),

$$\frac{\partial \{s\}}{\partial t} = -\tilde{F}(s) + \left\{\frac{\dot{q}}{T}\right\} + \left\{\frac{\dot{\phi}_{\text{irr}}}{T}\right\}, \qquad (2.74)$$

where $\tilde{F}(s) = -F_z(s)$, from equation (2.73).

Consider equation (2.74) as it applies to the combined atmosphere-ocean system (for which $\tilde{F}(s) = 0$), and let an overbar represent an average over a period long enough to smooth out natural variations. Over such a period, all state functions are constant, $\overline{\frac{\partial s}{\partial t}} = 0$, and,

$$-\overline{\left\{\frac{\dot{q}}{T}\right\}}_{oc+atm} = \overline{\left\{\frac{\dot{\phi}_{\text{irr}}}{T}\right\}}_{oc+atm}. \qquad (2.75)$$

The right side of equation (2.75) is positive definite because dissipation is an irreversible process. It follows that radiative transfer must *decrease* the entropy of the atmosphere-ocean system. General circulation climate models should obey equation (2.75), but it is not clear that most do so.

2.4.3 Carnot efficiency

According to Figure 2.2(c) the atmosphere has a net gain of radiative energy from latitudes 0° to 35°, and a net loss from latitudes 35° to 90°. If we use climatological mean temperatures (Appendix B) and weight appropriately with surface areas and with pressure, the difference of temperature between source and sink is ~ 15 K. The source and sink are separated in altitude as well as latitude. Solar heating takes place close to the ground, while thermal emission to space takes place from an emission level about 5 km above the surface (see §5.1.2 for a discussion of these matters). The vertical difference in temperature between source and sink is ~ 25 K. Vertical and horizontal differences together suggest a source-sink temperature difference of ~ 40 K.

For a mean atmospheric temperature of 250 K these data suggest a Carnot efficiency $\approx 40/250 = 0.16$. Investigators have contrasted this figure with the fraction of the solar energy that goes into kinetic energy, ≈ 0.01, according to Figure 2.2. The comparison is not appropriate. To the extent that the atmosphere is in a steady state, it does not produce any *net* work, and has zero efficiency. We may model the atmosphere better by a Carnot engine dissipating all of its mechanical energy in friction and returning this work partly to the source and partly to the sink in the form of heat (see Figure 2.5).

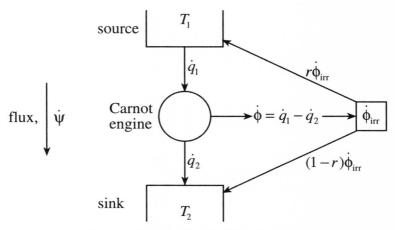

Figure 2.5 A steady-state atmospheric model.

Let $\dot{q}_1$ be the rate of heat loss by the source, let $\dot{q}_2$ be the rate of gain of heat by the sink, and let r be the fraction of heat dissipated that is returned to the source. The heat flux from source to sink is

$$\dot{\psi} = \dot{q}_1 - r\dot{\phi}_{\text{irr}}, \qquad (2.76)$$

where

$$\dot{\phi}_{\text{irr}} = \dot{q}_1 - \dot{q}_2. \qquad (2.77)$$

From Carnot's theorem,

$$\frac{\dot{q}_1}{T_1} = \frac{\dot{q}_2}{T_2}. \qquad (2.78)$$

It follows that,

$$\frac{\dot{\phi}_{\text{irr}}}{\dot{\psi}} = \frac{T_1 - T_2}{T_1 + r(T_2 - T_1)} \approx \frac{T_1 - T_2}{T_1 \text{ or } T_2} \approx 0.16. \qquad (2.79)$$

We may estimate $\dot{\psi}$ from Figure 2.2. The maximum poleward flux of heat by motions in each hemisphere is 4.5×10^{15} W. An average dissipation of ~ 2.5 W m^{-2} multiplied by the surface area of a hemisphere leads to $\dot{\phi}_{\text{irr}}/\dot{\psi} \approx 0.14$. In the light of all the uncertainties involved, this is close to the Carnot efficiency, suggesting that the thermal transformations in the atmosphere are essentially reversible.

2.4.4 Available potential energy

A question that has been discussed by many climatologists is: Which state function provides the best diagnostic for understanding the maintenance of the general circulation? The early literature chose the total potential energy, $(e+\pi)$, because, in an isolated, hydrostatic system, it is directly related to the kinetic energy by equation (2.67). However, the entire amount of the total potential energy cannot be transformed into kinetic energy. To see why this is so, consider the analogous situation of the potential energy of an isolated particle. The potential energy of the particle at the surface is usually taken to be zero because of the practical consideration that it cannot fall further than that. We are only interested in potential energies in excess of this minimum value. Similarly, if we attempt to rearrange the atmosphere in isolation in order to release kinetic energy, we will find a minimum value of the total potential energy, $(e+\pi)_0$, when colder, denser air always lies below warmer, lighter air. This is the lowest state of potential energy and, unless there is outside intervention, only the difference $(e+\pi) - (e+\pi_0)$ is available for transformation into kinetic energy.

Unlike the particle analogy, $(e+\pi)_0$ is not unique, but depends upon the processes that take place in the course of the rearrangement of the atmosphere. If the processes are adiabatic, however, the final state is defined and, from a well-known theorem in thermodynamics, it yields the maximum conversion to mechanical energy. A dry adiabatic process may be performed as follows. Take independent volumes of air and spread them out into a thin layers having the same potential temperature. The layers are then ordered with the potential temperature increasing uniformly upwards, which ensures that each layer has denser air below it. This is the lowest state of total potential energy that can be achieved adiabatically. Lorenz called the difference $(e+\pi) - (e+\pi)_0$, calculated adiabatically, the *available potential energy*.

The available potential energy has been employed in many studies of transformations of energy in the atmosphere. Its shortcoming as a tool for this purpose is that it does not bring in the thermodynamic efficiency of the system, and refers to the maximum energy available, constrained only by geometric and energetic considerations.

To put the problem in another way, equations (2.62) and (2.65) provide no a priori reason why *all* of the available potential energy cannot end up as kinetic energy. The equations should contain a Carnot efficiency if they are to satisfy the second law of thermodynamics.

2.4.5 Availability

In classical thermodynamics, the *availability*, a, is introduced to assess the maximum work available from a system in a given state, contained within given surroundings. The rate equation for availability is, from equations

(G.7), (2.13), and (2.17),

$$\rho \frac{da}{dt} = \rho \left(\dot{q} + \dot{\phi}_{\text{irr}} \right) \left(1 - \frac{T_s}{T} \right) - \rho p \frac{dv}{dt} , \qquad (2.80)$$

where T_s is the temperature of the "surroundings."[7]

The balance equation for the inventory of availability plus potential energy is

$$\frac{\partial \{a + \pi\}}{\partial t} + \tilde{F}(a + \pi) = \left\{ \left(\dot{q} + \dot{\phi}_{\text{irr}} \right) \left(1 - \frac{T_s}{T} \right) \right\} + \left\{ \frac{\vec{u}}{\rho} \cdot \nabla_h p \right\} , \qquad (2.81)$$

where, from (G.7) and (2.21),

$$\tilde{F}(a + \pi) = - \int_{surface} \left[F_z(e + \pi) - \frac{T_s}{T_g} F_z(h + \pi) \right] dA . \qquad (2.82)$$

While it is not necessary to do so, we may use (2.71) and (2.72) in order to define T_s so that $\tilde{F}(a + \pi) = 0$.

If we compare equations (2.65) and (2.81), we see that the availability can provide the ageostrophic drive in the same way that the total potential energy does. By way of contrast with the total potential energy, the availability equation does provide a constraint on the amount of heat that can be transformed into kinetic energy by introducing the Carnot efficiency, $(1 - \frac{T_s}{T})$. The trend of modern atmospheric thermodynamics has been toward using the availability, or a similar state function, as a basis for analyzing energy transformations.

2.5 Reading

Atmospheric dynamics is discussed, at approximately the level of this book, by

Lindzen, R.S., 1990, *Dynamics in atmospheric physics*. Cambridge: Cambridge University Press.

The most useful text book on atmospheric thermodynamics in the English language is

Iribarne, J.W., and Godson, W.L., 1973, *Atmospheric thermodynamics*. Dordrecht: D. Reidel.

Some of the material in this chapter is also contained in

Houghton, J.T., 1986, *The physics of atmospheres*. London: Cambridge University Press.

---

[7] The word "surroundings" means an undefined reservoir with which the atmosphere may exchange heat. While this notion may be acceptable analytically, it is unsatisfactory physically.

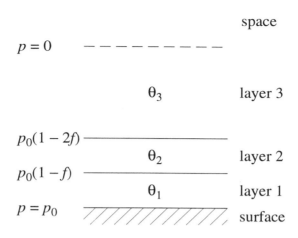

Figure 2.6 A three-layer atmosphere.

Houghton's book is recommended reading for many other topics in atmospheric physics. Recent compilations of climate data may be found in

Peixoto, J.P., and Oort, A., 1992, *Physics of climate*. New York: American Institute of Physics.

This book is a good choice for advanced reading on the climate.

2.6 Problems

Asterisks* and double asterisks** indicate higher degrees of difficulty.

2.1 **Margules' calculations.** At the turn of the century, the Austrian meteorologist Margules made a series of revealing calculations concerning transformations of energy in the atmosphere. The following calculations illustrate the nature of his investigations.

(i) As a first step, calculate the enthalpy or the total potential energy (see equation 2.58) per unit area for a dry atmospheric slab between pressures p_1 and p_2 that is hydrostatic and has a constant potential temperature. The base pressure of the atmosphere is p_0. Divide the slab enthalpy by the mass of the atmosphere to give the contribution of the slab to the mean specific enthalpy averaged over the entire atmospheric column.

(ii) Figure 2.6 shows an atmosphere consisting of three slabs. We calculate the change in mean specific enthalpy when the lower two slabs are interchanged. Exactly how this happens is not part of the problem, which is concerned only with the energetics. State the result in terms of the tem-

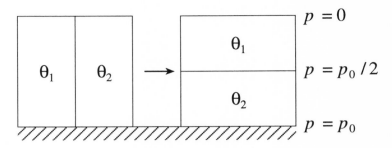

Figure 2.7 A two-column atmosphere.

perature difference $(\theta_1 - \theta_2)$. Set $f = 0.2$ and use c_p and r for dry air from Appendix A.

(iii) Perform a similar calculation for a two-column atmosphere changing to a two-slab atmosphere, as shown in Figure 2.7.

(iv) If the two atmospheres are completely isolated, the enthalpy cannot change by itself. Extend the argument of §2.3.5 or equation (2.15) to show that the kinetic energy must somehow change. If $(\theta_1 - \theta_2) = 10$ K, and if the atmosphere starts from rest, what will be the velocities after the changes for each atmosphere?

2.2 The tephigram. Even in the age of computers, graphical methods for handling ascent data can be useful if they can convey physical information. One graphical method of representing atmospheric data (*aerological diagram*) is the *tephigram*, in which the orthogonal axes are temperature, T, and $\ln \theta_d = s_{\text{dry}}/c_{p,a}$, see Figure 2.8 (the name *tephigram* comes from the symbol ϕ that was commonly used for entropy a decade or two ago).

(i) Show that the area under an ascent curve is

$$\int T \, d\ln \theta_d = \frac{q_{\text{dry}}}{c_{p,a}},$$

where q_{dry} is the heat interaction with the dry component of the air, considered as a separate system.

(ii) In the troposphere temperature decreases uniformly with height (on the average), and the axes of Figure 2.1 could be relabelled to produce a tephigram. The trajectory shown is adiabatic, but there is an area under the curve. Show that, in the limit $l \gg r_v T$, the area is,

$$\int_B^C T \, d\ln \theta_d = \frac{l}{c_{p,a}} \{m_v(B) - m_v(C)\}.$$

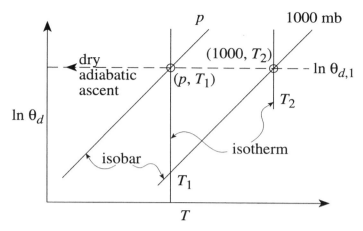

Figure 2.8 The ascent of dry air represented on a tephigram. The isobars are not straight lines, nor are they evenly spaced. They are included so that the state of a dry parcel can be plotted without having to perform a preliminary calculation.

(iii) A parcel of dry air undergoes a reversible lifting process that takes it from $(T_1 = 10°C, p_1 = 10^5$ Pa$)$ to $(T_2 = 0°C, p_2 = 8.5 \times 10^4$ Pa$)$. The trajectory may be described by a straight line on the tephigram between these two points. What are the changes in: internal energy; enthalpy; entropy? How much heat has been supplied to the parcel and how much work was performed?

2.3 **Moist and dry adiabats.** A parcel of air rises adiabatically from $p_0 = 10^5$ Pa, $T_0 = 288$ K by one of three different trajectories:

I. In equilibrium with a dry adiabatic and hydrostatic environment.

II. In equilibrium with a moist adiabatic and hydrostatic environment.

III. Corresponding to a certain extent with the cumulus core discussed in §5.1.3. Ascent follows a moist adiabat, but in pressure equilibrium with a dry adiabatic environment with the same surface temperature.

(i) Write expressions for the dry potential temperature θ_d for each trajectory in terms of the temperature and the saturated mixing ratio of the parcel. Use the approximations (2.30) and (2.32). How do trajectories II and III differ?

(ii) What is the saturated mixing ratio of water at the bottom of the trajectory? If we neglect the mixing ratio of water vapor at the top of the trajectory, what is the asymptotic difference of dry potential temperature between the parcel and its environment for trajectory III?

2.4 **Moist entropy.** A 1 kg parcel of saturated air at a pressure of 10^5 Pa and a temperature $T_1 = 20°C$ (containing no condensed water initially) is cooled adiabatically at constant pressure to $T_2 = 0°C$. The condensed liquid is retained in the parcel. Entropy changes are calculated using a variety of approximations. The purpose of this question is to evaluate some of them.

Calculate the water vapor-mixing ratio and the latent heat of evaporation for 20°C and 0°C. Use equation (2.27) to calculate the decrease in entropy as precisely as possible, and then calculate it with the following three approximations:

(i) $l(T) = l(0°C)$;

(ii) $c = 0$;

(iii) $m = 0$ but $m_v \neq 0$. Treat the specific heats as constants between 0°C and 20°C.

2.5 **Specific heat of saturated vapor.** In the following problem we isolate the effect of evaporation or condensation on the rate of change of temperature in an atmospheric slab, assuming all other factors to be constant. The slab lies between two horizontal, constant-pressure surfaces and is isolated except for a vertical change of the radiation flux, ΔF. The pressure does not change and there are no motions.

(i) Using the approximation to the enthalpy, equation (2.53), shows that, for a column of unit area,

$$\frac{\partial \{h\}}{\partial t} = -\Delta F = \left\{ \left(c_p + \frac{l^2 m_v}{r_v T^2} \right) \frac{\partial T}{\partial t} \right\}.$$

(ii) The upper and lower pressures for the slab are 9.5×10^4 Pa and 10^5 Pa, respectively, and the initial temperature is 20°C. The flux divergence over the layer is 50 W m^{-2}. What is the initial rate of change of temperature assuming:

(a) that the layer is dry;

(b) that the layer is in contact with a liquid surface and is saturated at all times.

Neglect the effect of the water-vapor pressure on the total pressure, and assume that the temperature change is the same at all points within the slab.

2.6 **A Föhn wind.*** A wind blows from ground level, over a mountain, cooling and precipitating water, and descending to ground level again as a hot, dry, Föhn wind (see Figure 2.9). Conditions measured at three of the points are given in Table 2.3. The flow is adiabatic and all condensed

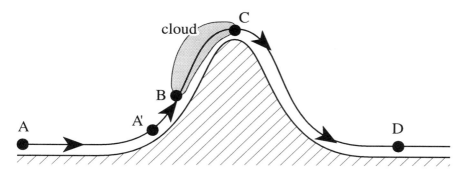

Figure 2.9 A Föhn wind

Table 2.3 Thermodynamic data

	p, Pa	m	T,° C
D	10^5	4×10^{-3}	38.0
A	10^5	10^{-2}	21.5
A'	8×10^4	5×10^{-3}	5.0

moisture is immediately precipitated. Although they are shown on the same streamline, the points A and A' represent two different flows, both of which end up at D.

(i) What are the dry potential temperatures at A, A', and D?

(ii) The dry potential temperature and the water vapor-mixing ratio are constant over the noncondensing portions of the trajectories. Show that the condensation temperatures, T_c, at B and C are given approximately by the relation:

$$\ln \frac{m_0}{m} + \frac{l}{r_v}\left(\frac{1}{T_0} - \frac{1}{T_c}\right) + \frac{c_{p,a}}{r_a} \ln \frac{\theta_d}{T_c} = 0,$$

where $m_0 = \frac{p_c(T_0)}{p_0}\frac{r_a}{r_v}$ is the mixing ratio for air at $p_0 = 10^5$ Pa saturated at $T_0 = 0°$C.

Calculate T_c at points B and C. To do so on a hand calculator approximate $\ln \frac{\theta_d}{T}$ by $\frac{\theta_d - T}{T}$. The maximum errors involved with this approximation are less than 1 K in T_c and 0.1 K in θ_e. Also assume that all coefficients are constant at their values at 0°C.

(iii) Calculate the equivalent potential temperatures for A, A', and D. Is A or A' the most likely origin for the air found at D?

2.7 **Altitude calculation.*** In Problem 2.6, the flow is embedded within, and in pressure equilibrium with, a hydrostatic air mass having a ground temperature of $T_0 = 21.5°C$ and a constant lapse rate, $\Gamma_{\text{obs}} = 6.5 \text{ K km}^{-1}$. This means that the flow itself is *not* hydrostatic.

(i) Show that the height, z, and the temperature, T, in the flow are related by

$$\ln\left(1 - \frac{\Gamma_{\text{obs}} z}{T_0}\right) = -\frac{\Gamma_{\text{obs}}}{\Gamma_{\text{dry}}} \ln \frac{\theta_d}{T}.$$

(ii) What is the height of the mountain, and what is the thickness of the cloud for flow A?

2.8 **Mixing of saturated air masses.*** Two equal masses of air are both saturated, but at slightly different temperatures, T_1 and T_2. The two masses are mixed together adiabatically without changing the potential energy, and the final temperature is T.

(i) Obtain an expression for the difference between the final temperature and the mean of the two initial temperatures. To find a solution, expand the enthalpy to second order in $(T_{1,2} - T)$, and assume that l and c_p do not depend upon temperature. Show that, to this accuracy,

$$T_1 + T_2 - 2T = \frac{-\alpha + \sqrt{\alpha^2 - 4(T_1 - T_2)^2}}{2},$$

where

$$\alpha \approx \frac{4T(l/r_v T + c_p T/m_v l)}{(l/r_v T)^2 - 2l/r_v T}.$$

The expression for α is obtained by assuming $m_v \ll 1$ and using the approximate form of the Clausius-Clapeyron equation, (G.26).

(ii) Does water condense or evaporate to maintain saturation? With $T = 250$ K, $m_v = 4 \times 10^{-3}$, and $T_1 - T_2 = 10$ K, calculate to first order the temperature difference and the fraction of the vapor that changes phase.

2.9 **Available potential energy.**** Figure 2.10 shows the same atmospheric data displayed in three different ways. (a) shows climatological mean temperatures for the Northern Hemisphere winter. (b) shows the same data in the form of dry potential temperatures; note that the potential temperature increases with height, everywhere, and the atmosphere is, therefore, stable for dry processes, everywhere. (c) shows the same data, rearranged dry adiabatically, that is, conserving dry potential temperature: The rearrangement has placed the lowest potential temperatures as far down as possible and the highest potential temperatures as far up as possible. This rearrangement places each parcel of air above air that is more dense and below air that is less dense. It requires that potential isotherms

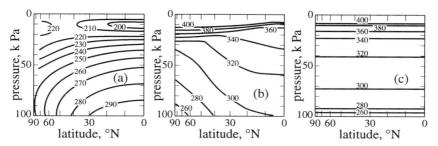

Figure 2.10 Available potential energy. (a) Climatological mean temperatures for the northern winter. (b) As for (a), but dry potential temperatures. (c) Dry potential temperatures in the minimum enthalpy state

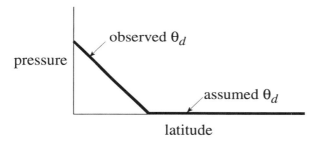

Figure 2.11 θ_d convention.

are horizontal. (c) is the lowest enthalpy state that can be reached adiabatically. The following steps are required to calculate the available potential energy.

(i) Show that the pressure of a θ_d isotherm in Figure 2.10, panel (c) is given by

$$\tilde{p}(\theta_d) = \frac{1}{A} \int_A p(\theta_d, \lambda) \, dA \;,$$

where $p(\theta_d, \lambda)$ is the pressure along a θ_d isotherm prior to adiabatic rearrangement, λ is the latitude, and dA is a small element of area parallel to the surface. For this relation to be correct it is necessary to observe the convention that an isotherm is continued along $p = p_0$ at all latitudes where it does not exist, see Figure 2.11.

(ii) Show that the enthalpy for a vertical column of unit cross section is

$$\{h_{\text{dry}}\} = \frac{c_p}{\left(1 + \frac{r}{c_p}\right) g p_0^{r/c_p}} \int_0^\infty p^{1+r/c_p}(\theta_d) \, d\theta_d.$$

The convention of Figure 2.11 is again used.

(iii) Hence show that the available potential energy per unit area of surface is

$$\{h_a\} = \frac{c_p}{\left(1 + \frac{r}{c_p}\right) g p_0^{r/c_p}} \int_0^\infty \left\{\widetilde{p^{1+r/c_p}}(\theta_d) - \tilde{p}^{1+r/c_p}(\theta_d)\right\} d\theta_d,$$

where the tilde operator was defined in (i).

2.10 **Solar energy.**** A flux of solar energy, $F_s(e)$, is absorbed by a solid, black flat surface that maintains equilibrium by emitting the same flux of radiation, but as low temperature, planetary radiation at a temperature, $T_e \approx 250$ K. The temperature of the sun is $T_s \approx 6000$ K.

The flux of entropy on a radiation field is

$$F(s) = \frac{4}{3} \frac{F(e)}{T},$$

where T is the emission temperature, here either T_s or T_e.

(i) According to the discussion of §2.4.5, the availability function defines the maximum work available from the solar energy. If we assume that any atmospheric heat engine must use the emission temperature as the temperature of the "surroundings," show that the maximum efficiency of conversion of solar energy into mechanical work is,

$$\eta = \frac{da}{de} = 1 - \frac{4}{3}\frac{T_e}{T_s} \sim 94.4\%.$$

(ii) Some solar energy systems obtain work from solar radiation by heating water by about 50 K, with a resulting Carnot efficiency of $\approx 20\%$. Why is this figure lower than that obtained in (i), and what must be done to avoid such low efficiencies?

CHAPTER 3

ABSORPTION AND SCATTERING

Atmospheric studies require a clear understanding of the processes that take place when a solar photon or a thermal photon interacts with a molecule or with an aerosol particle. These processes may be called absorption *or* scattering, *although the distinction is one of emphasis rather than fundamentals.*

Electronic absorption spectra of gases are important for photochemistry, and are relatively straightforward to handle numerically. Absorption spectra caused by vibration-rotation transitions are, on the other hand, too complicated for straightforward numerical treatment. There are hundreds of thousands of vibration-rotation lines of 30 or more different atmospheric species that we must consider. Digital archives of these data are now available and form the basis for modern theoretical treatments of atmospheric absorption.

The shape of each individual line is also required, but most of this information is not in the archives and must be provided. Our knowledge of line shapes at low pressures is based on Doppler shifts for isolated molecules. For colliding molecules, simple theories can account for the effect of molecular collisions on the centers of lines. When far line wings matter, the theoretical problem is more difficult, but semi-empirical treatments are available. From all of these data and theories, numerical calculations for gaseous atmospheres can be made with an accuracy that is generally acceptable for meteorological and climate calculations.

Particle interactions with radiation are complicated by the natural variability of shapes and sizes of aerosol particles. Mie's theory offers a complete numerical treatment of extinction by a dielectric sphere and is fundamental to our understanding of scattering in the atmosphere. Asymptotic analytical theories for large and small particles, based on accepted approximations of physical optics, are also of value.

From these theories we have reached a general understanding of particle extinction, the angular distribution of scattered radiation, and the role of polarization. A crucial parameter is the ratio of the circumference of a particle to the wavelength of light. When this parameter is small, as for a molecule, the angular distribution of scattered light is relatively simple, scattering depends on the wavelength, and polarization is strong; when the parameter is large, as for a cloud droplet, the angular distribution is complicated, polarization is relatively unimportant, and the color of scattered light is white.

Water vapor can exist in either the vapor phase or in a condensed phase. The effect on scattered light of a given number of molecules is maximized when they are aggregated into particles of sizes comparable to the wavelength of light.

3.1 Interaction coefficients and cross sections

In §1.5.1 we distinguished between two possible ends for the energy of a photon interacting with a molecule: The energy could go into the thermal reservoir or a photon could be re-emitted. In the terminology of *radiative transfer theory* (Chapter 4), these processes are *absorption* and *scattering*, respectively. On a probabilistic basis, both can occur simultaneously: The joint process is called *extinction*, and the probability of scattering is called the *single-scattering albedo*.

Numerical coefficients for extinction, absorption, and scattering are derived experimentally using *Lambert's law*,

$$dI_\nu = -e_\nu n I_\nu \, dl \;, \tag{3.1}$$

where I_ν is the *radiance* or *radiant intensity*,[1] dI_ν is the *gain* in radiance when traversing the infinitesimal path dl, and n is the number density of absorbing molecules. e_ν is, by definition, the *extinction coefficient per molecule*.

If extinction acts alone, that is, in the absence of emission, (3.1) may be integrated,

$$I_\nu(l) = I_\nu(0) \exp\left(-\int_0^l e_\nu(l') n(l') \, dl'\right) \;. \tag{3.2}$$

The *transmission*, $\mathcal{T}_\nu$, and the *absorption*, $\mathcal{A}_\nu$, are defined by

$$\mathcal{T}_\nu(l) = 1 - \mathcal{A}_\nu(l) = \frac{I_\nu(l)}{I_\nu(0)} \;. \tag{3.3}$$

The exponent in equation (3.2) is defined in §4.2.1 to be the *optical path*.

---

[1] The radiance is defined in §4.1; for the present purposes it may be thought of as a collimated flux of radiant energy.

Equation (3.1) may also be applied either to absorption or to scattering, but with the *absorption coefficient*, k_ν, or the *scattering coefficient*, s_ν, respectively, in place of e_ν. Since extinction is the combination of absorption and emission,

$$e_\nu = k_\nu + s_\nu \, , \tag{3.4}$$

and the probability of scattering, or the single-scattering albedo is,

$$a_\nu = \frac{s_\nu}{e_\nu} \, . \tag{3.5}$$

According to the conventions of chemical kinetics, the extinction process may be seen as a collision between a photon of negligible size and a molecule having a *collision cross section* for the process. As defined by equation (3.1), e_ν is this cross section.[2] Since, on the average, a photon can collide ne_ν times in unit distance, the *mean free path* of a photon is,

$$\lambda_\nu = (e_\nu n)^{-1} \, . \tag{3.6}$$

In the following sections we consider absorption and scattering, first by gases and then by aerosols, including water droplets. While it is convenient to treat gases and aerosols separately, they too are related. The most important distinction between them is the state of aggregation of the molecules. Aggregates do not have outer shell electrons that can make quantized transitions, leading to *line spectra*, as is possible for isolated molecules. Instead, aggregates scatter more than the individual molecules contained in them, and they exhibit *continuum absorption* due to the collective behavior of outer shell electrons.

Because of these differences, the focus of studies differs for gases and aerosols. Scattering from gaseous molecules (*Rayleigh scattering*) is of relatively small importance. The major topics for gases are: the electronic bands, and their relevance to photochemistry; the many vibration-rotation bands of minor species; and the shapes of their individual lines. Data on atmospheric molecules that are needed for atmospheric calculations are now largely available.

For aerosols, attention focuses on the scattering process. Because the scattered photon retains a memory of the vector properties of the incident photon, scattering can be very complicated. Studies focus on the angular distribution, the polarization, the mathematical difficulty of scattering calculations (see §4.2.2), and the natural variabilities of aerosol particle shapes and number densities.

The line between absorption and scattering can be crossed. In §4.1.5 we discuss a practical case for which absorption can turn into scattering (*resonant scattering*) if the pressure is low enough. Such transitional cases are of little importance in the middle and lower atmospheres.

---

[2]Note the dimensions, $n \sim (length)^{-3}, l \sim (length), e_\nu \sim (length)^2$.

3.2 Absorption by gases

3.2.1 Lines and continua

An overview of the principal atmospheric absorption bands is provided by Figure 1.3, panels (b) and (c). Most of these bands have a fine structure that is not shown because the spectral resolution is insufficient. There are many bands, amongst them bands of O_2, O_3, H_2O, HDO, CO_2, CH_4, and N_2O. Weaker bands, which are not shown in this figure, can also play a part in atmospheric calculations.

Absorption bands are classified according to the nature of the principal quantum transition. The nature of a transition is loosely coupled to the magnitude of the energy change, but with some ambiguities: *Electronic transitions* involve the largest energies; *rotational transitions* involve the smallest; and *vibrational transitions* lie between. From Planck's relation[3] electronic bands occupy the shortest wavelengths, less than 1 μm, rotation bands occupy wavelengths longer than 20 μm, and vibration bands lie between. Because of the hierarchy of energies, vibration bands also have rotational structure (*vibration-rotation band*) and electronic bands can have both vibrational and rotational fine structure.

Bands of O_2 and O_3 at wavelengths less than 1 μm are electronic transitions. Absorption cross sections for the principal electronic bands of oxygen and ozone are illustrated in Figures 3.1 and 3.2. The peak cross sections, $\sim 10^{-21}$ m^2, are extraordinarily large, and are more characteristic of a metal than a gas. Despite the small amount of ozone in the atmosphere, no solar radiation penetrates to the ground at wavelengths less than ~ 310 nm. These absorption bands are relatively uncomplicated continua, for the reason that all absorptions result in dissociation of the molecule, so that the upper state is not quantized (see §6.2 for a discussion of oxygen and ozone photolysis). Pressure and temperature have some effect upon them but, for the most part, this creates no difficulties for numerical calculations.

Not all electronic absorptions lead to dissociation. Of some importance are the "red" and "infrared" bands of molecular oxygen, formed by transitions in the triplet ground electronic state of the molecule. There are several bands, each associated with a vibrational transition. Their appearance is similar to that of a vibration-rotation band with an abnormally high vibrational frequency, and they must be treated similarly.

Vibration-rotation and pure rotation bands consist of rotation lines closely spaced about the frequency of a vibrational transition, or zero frequency in the case of rotation bands. To illustrate the detail present in a

---

[3] Energy is proportional to frequency, for which reason frequency is more important than wavelength for discussing spectra; the *wavenumber* (unit cm^{-1}; see Appendix A) is often used to identify vibration and rotation bands. For reference: 1 μm $\equiv$ 10 000 cm^{-1}; 10 μm $\equiv$ 1000 cm^{-1}; 100 μm $\equiv$ 100 cm^{-1}.

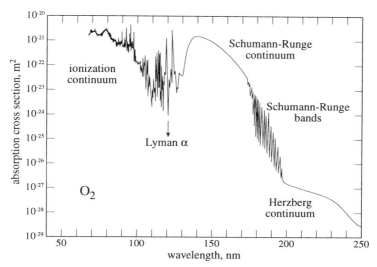

Figure 3.1 Electronic bands of oxygen from 50 to 250 nm. The identified spectral regions play different roles in the photolysis of oxygen, see Chapter 5. Hydrogen Lyman-α is a strong emission line in the solar spectrum that fortuitously falls in a narrow window in the oxygen absorption spectrum.

vibration-rotation band, Figure 3.3 shows a small section of one methane band at two spectral resolutions.

To a first approximation, atmospheric molecules behave as if isolated from one another. At a pressure of 1 bar, molecules spend 50 to 100 times more time in free flight between collisions than they do in close proximity to each other. At lower pressures this figure is proportionately larger. Thus the frequencies of lines, and the general appearance of a vibration-rotation band, can be calculated with precision from the energy levels of an isolated molecule.

The same theory of isolated molecules also allows *line strengths*, or integrated line absorption coefficients, to be calculated from dipole moment information:

$$S_{line} = \int_{line} k_\nu \, d\nu \ . \tag{3.7}$$

Most molecules that affect atmospheric thermodynamics are relatively simple, and the necessary molecular parameters for quantum-mechanical calculations are known. Although it is a formidable task, the position and strength of every known atmospheric absorption line has been calculated from high-order perturbation theory, and the results stored in one of a number of digital archives. These archives presently (in 1994) list almost one million atmospheric absorption lines, of which 300,000, grouped in 1000

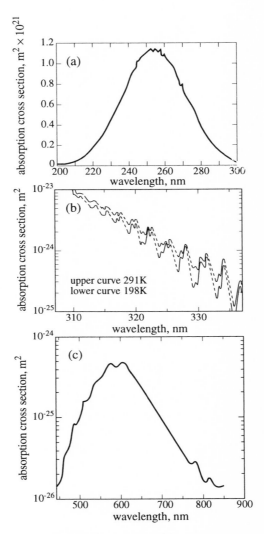

Figure 3.2 The electronic bands of ozone. (a) The Hartley bands, (b) the Huggins bands, (c) the Chappuis bands. The peak atmospheric absorption in the Chappuis bands is only a few percent, but since it occurs near to the maximum radiance of the sun, this band absorbs important amounts of solar energy.

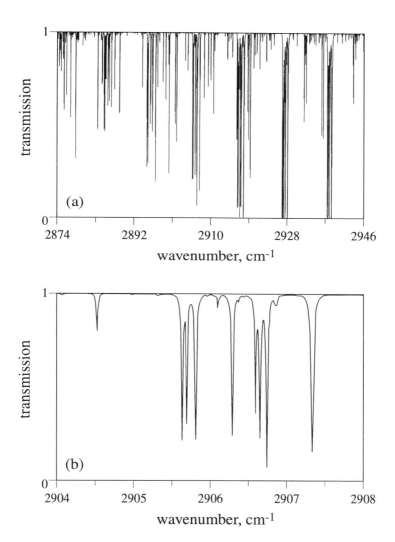

Figure 3.3 Absorption spectrum of the 3.44 μm vibration-rotation methane band. (a) 2874–2946 cm^{-1}, (b) 2904–2908 cm^{-1}. These are theoretical spectra, calculated from a digital archive. The lower spectrum is almost fully resolved. The absorption path is the entire atmosphere above 10 km.

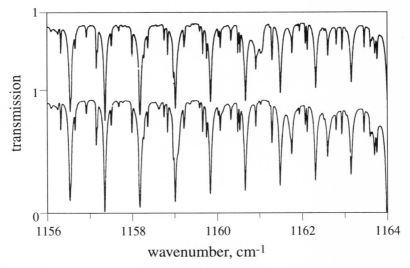

Figure 3.4 Observed and theoretical atmospheric absorption spectra. The upper spectrum is observed while the lower is calculated from a digital archive. This spectral region contains both ozone and nitrous oxide lines.

bands, may have some importance for the thermal state of the atmosphere. The precision of these data are illustrated by the comparison between theoretical and observed spectra of nitrous oxide and ozone shown in Figure 3.4. It requires careful comparison to establish that there are any differences at all between these spectra.

The availability of archives in digital formats that are compatible with large computers has greatly simplified the task of atmospheric calculations. It is now normal practice to store an archive in memory and to call for whatever spectral data may be required. Attention need be given to two points. First, the archives give line strengths for a single temperature. Line strengths are proportional to the population of the lower state of the transition, and the archives give the partition functions and lower state energy-level information needed to calculate the state populations from Boltzmann's law (equation 1.7). Temperature variations of line strengths can be of great importance, but the problem is theoretically tractable. Second, the archives give some information about the shapes of lines, but this information does not have the rigor of the position and strength information. The user must supply line shape information, and the quality of his or her results may depend importantly upon this judgement.

3.2.2 Line shapes

Absorption coefficients are related to the line strength by,

$$k_\nu = S_{line} f(\nu - \nu_0), \qquad (3.8)$$

where

$$\int_{line} f(\nu - \nu_0) \, d\nu = 1. \qquad (3.9)$$

ν_0 is the frequency of the quantum-mechanical transition, and $f(\nu - \nu_0)$ is called the *line shape*.

There are two circumstances under which the line shape takes a particularly simple form. The first is for low pressures, when collisions are rare, and the molecule behaves as if isolated. The second applies to higher pressures when collisions are the main factor in line broadening.

For an isolated molecule the only line broadening mechanism is the Doppler shift associated with molecular motions. For thermal equilibrium between translational levels, the probability that one velocity component lies between v and $v + dv$, $p(v) \, dv$, is given by the Maxwellian distribution,

$$p(v) = \left(\frac{m}{2\pi \mathbf{k} T}\right)^{\frac{1}{2}} \exp\left(-\frac{mv^2}{2\mathbf{k}T}\right). \qquad (3.10)$$

If $v/\mathbf{c} \ll 1$, where $\mathbf{c}$ is the velocity of light, the Doppler shift for an emitted quantum with rest-frame frequency, ν_0, is

$$(\nu - \nu_0) = \frac{\nu_0 v}{\mathbf{c}}. \qquad (3.11)$$

If we eliminate v between (3.10) and (3.11), and normalize according to (3.9), we obtain the *Doppler line shape*,

$$f_D(\nu - \nu_0) = \frac{1}{\pi^{\frac{1}{2}} \alpha_D} \exp - \left(\frac{\nu - \nu_0}{\alpha_D}\right)^2, \qquad (3.12)$$

where

$$\alpha_D = \frac{\nu_0}{\mathbf{c}} \left(\frac{2\mathbf{k}T}{m}\right)^{\frac{1}{2}}, \qquad (3.13)$$

is the *Doppler line width*. For a water-vapor line at 200 cm^{-1} and a temperature of 300 K, the Doppler line width is 3.4×10^{-4} cm^{-1}.

In the lower atmosphere the pressure is such that the perturbation of the emission process by and during a collision causes the line to have a finite width. The interaction during the collision can be treated only with difficulty. Fortunately, a simple model developed by Michelson and Lorentz gives a useful approximation to the shape of a line in the vicinity of the line center.

Michelson and Lorentz made the assumption that the time spent during collision is so short as to be unimportant, and that the effect of the collision is simply to interrupt randomly the emission or absorption process. In classical terms, this replaces infinite, monochromatic wave trains by wave trains of finite duration. A Fourier analysis of a wave train of finite duration yields a spread of frequencies on each side of the original frequency.

Let the unperturbed wave be $e^{2\pi i t' \nu_0}$, where ν_0 is the unperturbed frequency. If the wave only exists between collisions, $0 < t' \le t$, the spectral amplitude at frequency ν is obtained from the Fourier integral,

$$\psi(\nu) = \frac{1}{2\pi} \int_0^t e^{2\pi i t' (\nu - \nu_0)} \, dt' \ . \tag{3.14}$$

The spectral radiance is proportional to the Poynting vector, or to the square of the spectral amplitude, $\psi(\nu)\psi^*(\nu)$, where the asterisk indicates the complex conjugate.

Times between collisions are assumed to follow a Poisson probability distribution,

$$p(t) = \frac{1}{\tau} \exp\left(-\frac{t}{\tau}\right) \ , \tag{3.15}$$

where τ is the mean time between collisions. The average spectral amplitude is then,

$$\begin{aligned}\overline{\psi(\nu)\psi^*(\nu)} &= \int_0^\infty \psi(\nu)\psi^*(\nu) p(t) \, dt \ , \\ &= \frac{2}{(2\pi)^4} \frac{1}{(\nu - \nu_0)^2 + \alpha_L^2} \ , \end{aligned} \tag{3.16}$$

where, from elementary kinetic theory,

$$\alpha_L = \frac{1}{2\pi\tau} = \sum_i n_i \sigma_i^2 \left[\frac{2\mathbf{k}T}{\pi}\left(\frac{1}{m} + \frac{1}{m_i}\right)\right]^{\frac{1}{2}} \ . \tag{3.17}$$

α_L is the *Michelson-Lorentz line width*; n_i is the number density of perturbing molecules of species; i, m_i, and m are molecular masses of perturbers and absorbers, and σ_i is the molecular collision diameter.

The line shape (3.16) has a maximum at $\nu = \nu_0$, but a more careful treatment of the Michelson-Lorentz problem (which we omit) shows that the center should, generally, be shifted by a fraction of the line width; the shift is zero for collisions between rigid spheres. Adding this feature, and applying the normalization (3.9), gives the *Michelson-Lorentz line shape*,

$$f_L(\nu - \nu_0) = \frac{\alpha_L}{\pi[(\nu - \nu_0 - A\alpha_L)^2 + \alpha_L^2]} \ . \tag{3.18}$$

ABSORPTION AND SCATTERING

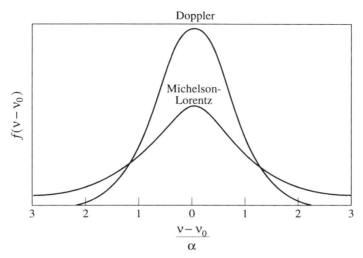

Figure 3.5 Michelson-Lorentz and Doppler line shapes. For the purpose of this comparison α_D and α_L are chosen to be equal.

Where known, Michelson-Lorentz widths and the *line-shift parameter*, A, are listed in the spectral archives.

Michelson-Lorentz and Doppler shapes are compared in Figure 3.5. The largest differences are in the wings of the lines, the importance of which is discussed in §3.2.3. Both line shapes have been confirmed in the laboratory under appropriate circumstances. Figure 3.6 shows a comparison between a measured line of water vapor and a calculation based on the Michelson-Lorentz theory. Over the observed frequency range, the agreement is very good.

From equation (3.17), the Michelson-Lorentz width is proportional to the number density of perturbing molecules. The important perturbers are normally oxygen and nitrogen molecules, and the width will then be proportional to the air pressure. A typical water-vapor line width with air as the perturber is 0.05 cm^{-1} at 1 bar (considerably narrower than the self-broadened width in Figure 3.6). For the 200 cm^{-1} water-vapor line that we introduced earlier, Michelson-Lorentz and Doppler widths are equal at a pressure of $\sim 10^3$ Pa, occurring at an altitude near to 30 km. Below this altitude the Doppler shape can usually be neglected. Above this level the Doppler shape may be important. A hybrid line shape, the *Voigt profile*, a convolution between the Michelson-Lorentz and the Doppler profiles, is available for the transition region, if needed.

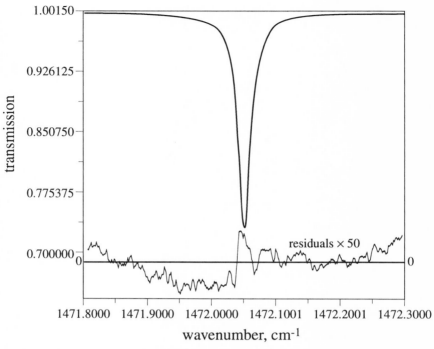

Figure 3.6 Observed and theoretical line shapes are compared. The transmission of a line at 1472.0520 cm^{-1} was measured in a 0.15 cm cell containing pure water vapor at a pressure of 2.175×10^3 Pa and a temperature of 297.65 K. This is compared with a calculation based on the Michelson-Lorentz line shape with a line width of 0.475 cm^{-1} at 1 bar. The line at the bottom shows the residuals multiplied by 50. For this comparison the collisions are between like molecules. This does not change the nature of the line-broadening mechanism, but it does give line widths that are much greater than those for the case of collisions between dissimilar molecules, for example, air and water vapor. Courtesy of R. May, Jet Propulsion Laboratory.

3.2.3 Line wings

Theoretical treatments of the detailed molecular interactions in close collisions indicate that the Michelson-Lorentz model should be satisfactory within ~ 2 cm^{-1} of the center of a rotation line. At larger displacements from the line center, however, the complex details of the interaction during the collision control the line shape. In the atmosphere, this displacement corresponds to a hundred or more line widths from the line center and, at first sight, it is puzzling that such displacements can be of any importance for atmospheric calculations, but they are. To understand this question better we may consider the data for the *water-vapor continuum* in Figure 3.7.

ABSORPTION AND SCATTERING

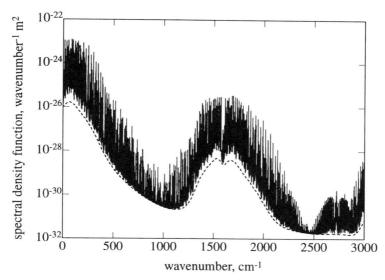

Figure 3.7 Absorption coefficients for water vapor. The "spectral density function" is the molecular absorption coefficient, divided by the factor $(\nu/c)\tanh(h\nu/kT)$, which is equal to ν/c for most frequencies that we shall discuss. The temperature is 296 K and the lines are broadened by 1.013 bar pressure of air. The full line is the complete absorption coefficient. The broken line is the "continuum contribution," calculated from contributions to the absorption coefficient 25 cm^{-1} or more from line centers. The difference between the full and broken lines is the "line contribution." The band at the left is the water-vapor rotation band, and the band in the center is the 6.3 μm vibration-rotation band of water vapor.

In order to construct Figure 3.7, the contribution to the absorption coefficient for each line is separated into that from displacements less than 25 cm^{-1} from line centers (*line contributions*), and that from displacements more than 25 cm^{-1} from line centers (*continuum contributions*). The continuum contribution is shown by the broken line. The total contribution is shown by the full line, and the line contribution (not fully resolved) is the difference between the two. In order to compare line and continuum contributions we must average over the lines. In the center of the major bands the line contributions outweigh the continuum, but in the windows near 1000 cm^{-1} and 2500 cm^{-1} the continuum dominates. These statements are more obvious when the high resolution spectra are available. From our earlier discussion, we may probably rely on the Michelson-Lorentz theory to calculate the line contribution, but it will have little significance for the continuum.

This leads to the next question: Why are these windows in the water-vapor spectrum of any importance? Figure 1.3, panel (b), shows the window near 1000 cm^{-1}. It has an ozone band in the center, but this does not significantly change the discussion. The point that may be garnered from Figure 1.3 is that radiation from the surface can escape directly to space through this window, exercising a strong and direct constraint over surface temperatures, and that the water continuum is the spectral feature that is most able to modulate this control. The water-vapor windows, and the far-wing shapes of water-vapor spectral lines are, in fact, crucial to climate calculations.

While line wing shapes raise questions of great complexity, it is fortunate that they have simple features. The frequency of a given interaction is, however complex, proportional to the rate of collisions. The absorption coefficient in the far wings of a line is, therefore, proportional to the collisional rate between absorbers and perturbers; that is, to the pressure. Because we are dealing with a continuum, its properties may be defined experimentally at a few frequencies and, given the known effect of pressure, a semi-empirical approach to line-wing absorption has proved to be effective.

3.3 Extinction by particles

The paradigm of scattering theory is the work of Mie, a *complete* solution for the interaction between a polarized plane wave and a dielectric sphere with complex refractive index. For a real refractive index only scattering can take place. For a complex index both scattering and absorption are possible and the coefficients depend upon both real and imaginary components.

Low- and medium-level clouds consist of water drops and, for these, *Mie's theory* is directly applicable. Air molecules are also spherical to a first approximation. Cirrus cloud particles and other solid aerosol particles are far from spherical. Nevertheless, some success has been had by treating nonspherical particles in terms of "equivalent spheres."

Despite the simple spherical geometry, Mie's theory is still remarkably complicated. The reason lies in the completeness of the solution, which combines in one expression all of the phenomena discussed in a textbook on wave optics: reflection, refraction, diffraction, absorption, interference, and polarization. The solution is presented in terms of infinite series that, by themselves, provide little physical insight, and that must be summed numerically. Only asymptotic forms of Mie's theory can be "understood" in the sense that wave optics can be "understood." In the following sections, after a brief summary of results from Mie theory, we consider two asymptotic forms, one for small and one for large particles, that together illustrate many of the important aspects of the theory.

3.3.1 Mie theory

The independent variables in Mie's theory are the real and imaginary parts of the refractive index of the sphere,[4] $\tilde{m} = \tilde{n} + i\tilde{n}'$, the scattering angles, and the *size parameter*,

$$x = \frac{2\pi r}{\lambda}, \qquad (3.19)$$

where r is the radius of the droplet and λ is the wavelength of light in the surrounding medium. In the case of a sphere, the relevant angles reduce to the *scattering angle*, θ, the angle between the directions of the incident and scattered photons (see Figure 3.10).

Figure 3.8 shows Mie theory calculations for three real refractive indices, and for values of x between 1 and 6. The quantities displayed are i_1 and i_2, two quantities that we do not need to define, except to note that they are the natural products of Mie theory, and that they are related to the Poynting vector, N, and the linear polarization, Π, of light scattered when the incident beam is unpolarized (e.g., incident sunlight) by

$$N = \left(\frac{\lambda}{2\pi d}\right)^2 \frac{(i_1 + i_2)N_0}{2}, \qquad (3.20)$$

and

$$\Pi = \frac{i_1 - i_2}{i_1 + i_2}, \qquad (3.21)$$

where the zero subscript indicates the incident Poynting vector, and d is the distance from the scattering sphere to the point of observation.

The calculations in Figure 3.8 are made at intervals of 10°. This is a sufficiently fine angular division to show most of the structure for $x = 1$ to 6. For large values of x, however, very complex angular fine structure can appear. The detail is very sensitive to the size parameter, x, and since the atmosphere contains a mixture of particle sizes, this fine structure may be smoothed out and is not usually observed.

The effect of an imaginary component to the refractive index is illustrated in Figure 3.9, where i_1 and i_2 are compared with and without an imaginary component. With minor exceptions, both parameters are smaller in the latter case, because they have been reduced by the radiation lost in absorption.

As has already been pointed out, Mie calculations, while correct, are difficult to describe and to understand. Rather than attempt this task we turn to two approximations, *Rayleigh's theory* for scattering by molecules,

---

[4] We distinguish between the optical properties of the sphere itself (with tilde) and the average properties of the medium containing air and widely separated spheres (without tilde). It is the latter that is required for radiative transfer theory, but the former is used in Mie calculations.

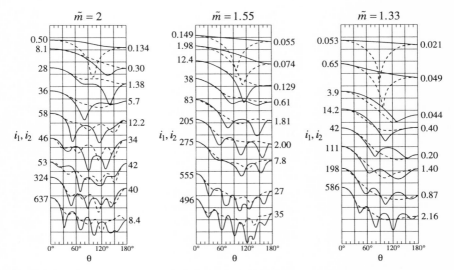

Figure 3.8 Scattering diagrams from Mie theory. The scales for i_1 (broken line) and i_2 (solid line) are logarithmic, with one division equal to a decade. The values for i_1 and i_2 at $\theta = 0°$ and $180°$ are shown at the sides. The different curves are for different values of x: From the top curve down, $x = 1$, 1.5, 2, 2.5, 3, 3.6, 4, 5, and 6.

which corresponds closely to Mie theory for $x \ll 1$, and van de Hulst's theory of *anomalous diffraction*, which corresponds to the limits $x \gg 1$ and $(\widetilde{m} - 1) \ll 1$.

3.3.2 Rayleigh's theory for small particles

The top three curves in Figure 3.8 are for $x = 1$. They exhibit similarities: For all three, i_1 varies little with the scattering angle, while i_2 has a sharp dip near $\theta = 90°$. This behavior is characteristic of calculations for $x \ll 1$ and is shared with Rayleigh's theory.

If a spherical particle is much smaller than the wavelength of light, we may assume the electric field of the incident wave, E_0, to be constant over the particle. The electric field will induce an electric dipole moment,

$$M^{(l,r)} = pE_0^{(l,r)}, \qquad (3.22)$$

where p is the *scalar polarizability* for the single particle. Both M and E_0 are oscillatory, with the time dependence $\exp(2\pi i c t/\lambda)$, where λ is the wavelength in the medium surrounding the particle. The indices (l, r) indicate linear polarizations, respectively, parallel or perpendicular to the plane containing the incident direction and the line of sight (the *plane of reference*), see Figure 3.10. This figure demonstrates the relation between

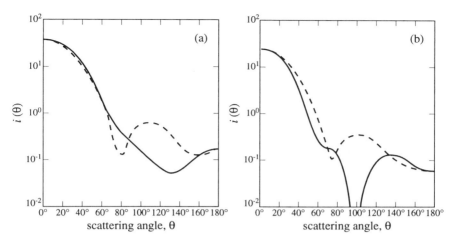

Figure 3.9 Mie calculations with and without an imaginary component to the refractive index. $i_1(\theta)$ is the broken line and $i_2(\theta)$ is the solid line. $x = 3$, $\tilde{n} = 1.315$. (a) $\tilde{n}' = 0.0$. (b) $\tilde{n}' = 0.4298$. (a) corresponds closely to $x = 3$ in the right-hand panel of Figure 3.8.

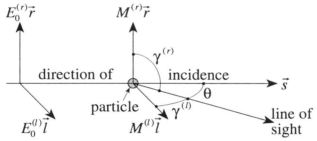

Figure 3.10 Dipole emission. $\vec{l}$ and $\vec{r}$ are unit vectors in the l- and r- directions. $\vec{s}$ is unit vector in the direction of propagation. $\vec{l}$ and $\vec{s}$ and the line of sight lie in the plane of reference. θ is the scattering angle, $\gamma^{(r)} = \frac{\pi}{2}$, and $\theta + \gamma^{(l)} = \frac{\pi}{2}$.

the *scattering angle*, θ, and the two angles, $\gamma^{(l,r)}$, between the line of sight and the direction of polarization.

The oscillating dipole, (3.22), emits a spherical wave with an electric field,

$$\begin{aligned} E^{(l,r)} &= \frac{1}{d\mathbf{c}^2} \left| \frac{\partial^2 M^{(l,r)}}{\partial t^2} \right| \sin \gamma^{(l,r)} , \\ &= \left(\frac{2\pi}{\lambda} \right)^2 \frac{|p|}{d} E_0^{(l,r)} \sin \gamma^{(l,r)} , \end{aligned} \quad (3.23)$$

where d is the distance from the dipole to the emitted wavefront. Equation (3.23) is the well-known solution of Hertz.

The relationship between the two angles θ and γ depends upon the direction of the polarization vector, as is illustrated in Figure 3.10. Two equations are, therefore, required to express equation (3.23) in terms of the scattering angle,

$$E^{(r)} = \left(\frac{2\pi}{\lambda}\right)^2 \frac{|p|}{d} E_0^{(r)}, \tag{3.24}$$

$$E^{(l)} = \left(\frac{2\pi}{\lambda}\right)^2 \frac{|p|}{d} E_0^{(l)} \cos\theta. \tag{3.25}$$

Equations (3.24) and (3.25) show that Rayleigh scattering is strongly polarized at right angles to the line of sight ($\theta = \frac{\pi}{2}$). Only perpendicular polarization (that is to say electric vector perpendicular to the plane of reference) is possible at this scattering angle, regardless of the polarization of the incident light. Polarization can be detected in light from the daytime sky by observing at right angles to the solar beam. The polarization is not complete, but that is because sky light has been scattered more than once (*multiple scattering*), while equations (3.24) and (3.25) describe a single scattering event. Multiple scattering will be discussed in Chapter 4.

The Poynting vector, N, has magnitude $|E|^2$, and is proportional to the radiance. If we are interested in the total energy flux and not in the polarization, we may add the Poynting vectors for the two polarizations. If the incident wave is unpolarized, with Poynting vector, N_0,

$$N_0 = N_0^{(r)} + N_0^{(l)} = 2N_0^{(r)} = 2N_0^{(l)}, \tag{3.26}$$

and, squaring and adding (3.24) and (3.25),

$$N = \left(\frac{2\pi}{\lambda}\right)^4 \frac{|p|^2}{d^2} N_0 \frac{1+\cos^2\theta}{2}. \tag{3.27}$$

Mie's and Rayleigh's theories are derived from electromagnetic theory, and the connection to radiative transfer theory is not obvious. We may show a connection between equations (3.1) and (3.27) by putting (3.1) in a form applicable to a single scattering particle, by writing $n\,dl = 1$ and $e_\nu = s_\nu$. Further, from the definitions of radiance and Poynting vector, the energy flux per unit area is proportional either to I or to N_0. Equation (3.1) may now be written,

$$-dN_0 = s_\nu N_0. \tag{3.28}$$

The energy loss $-dN_0$ may be identified with the total outflow of scattered radiation in all directions over the surface of a sphere of radius, d. An infinitesimal area on the surface of the sphere is $d\alpha = d^2 d\omega$, while

ABSORPTION AND SCATTERING

$d\omega = 2\pi d(\cos\theta)$; see Figure 4.3 for analogous geometry. Combining these results,

$$-dN_0 = \int_{sphere} N\,d\alpha = \int_{-1}^{+1} 2\pi N d^2\, d(\cos\theta)\,. \tag{3.29}$$

Substituting (3.27) into (3.29), integrating, and comparing to (3.28),

$$s_\lambda = \frac{8\pi}{3}\left(\frac{2\pi}{\lambda}\right)^4 |p|^2\,. \tag{3.30}$$

Equation (3.30) is the required connection between the two protocols, radiative transfer and electromagnetic theory, for the case of Rayleigh scattering.

We now specialize our results to the most important case of small particle scattering, namely, scattering by air molecules. To a first approximation air molecules may be treated as very small spheres. To do so, we take advantage of a well-known result from electromagnetic theory that enables us to express the bulk refractive index, m, of a gas consisting of an ensemble of molecules in terms of the polarizability, p, with the result,

$$s_\lambda = \frac{32\pi^3}{3\lambda^4}\left(\frac{m-1}{n}\right)^2\,. \tag{3.31}$$

Equation (3.31) represents a constant scattering coefficient per molecule because $(m-1) \propto n$.

The dominant feature of equation (3.31) is the dependence on the inverse fourth power of the wavelength; on this account alone, the scattering coefficient is 16 times larger at 400 nm in blue light than at 800 nm in red light. The refractive index of air also has a slight dependence upon wavelength, but this is a second-order effect. Some data for the atmosphere are given in Table 3.1. The final column is the total opacity of the atmosphere in a vertical direction, $\tau_1 = \int_0^\infty sn(z)\,dz$, where z is the altitude. In Chapter 4 this quantity will be called the *optical depth*; it is a measure of the importance of molecular scattering. If the optical depth is small compared to unity, as at the red end of the spectrum, scattering is relatively unimportant; but in blue and ultraviolet light molecular scattering is a major feature of atmospheric optics, and explains the blue color of the daytime sky.

To complete our discussion of Rayleigh scattering we must introduce an approach to the angular distribution of scattered light. This makes use of *phase function for scattering*, $P(\cos\theta)$, in terms of which the light scattered into a solid angle $d\omega$ is $P\frac{d\omega}{4\pi}$. From (3.28) and (3.29),

$$P\frac{d\omega}{4\pi} = \frac{N\,d\alpha}{\int_{sphere} N\,d\alpha} = \frac{Nd^2 d\omega}{s_\lambda N_0}\,. \tag{3.32}$$

Table 3.1 Data on molecular scattering by air

λ, nm	s_λ, m^2	$(m-1) \times 10^5$ †	$\tau_1^{\ddagger}$
200	3.551×10^{-29}	34.19	7.630
400	1.689×10^{-30}	29.83	0.363
600	3.202×10^{-31}	29.22	0.098
800	9.989×10^{-32}	29.01	0.021
1000	4.065×10^{-32}	28.92	0.009

†Refractive indices are evaluated at 0°C and 1 bar.
‡There are assumed to be 2.149×10^{29} molecules m^{-2} in an atmospheric column.

This relation is valid for any phase function, but specializing to Rayleigh scattering with (3.27) and (3.30),

$$P(\cos\theta) = \frac{3}{4}(1 + \cos^2\theta). \tag{3.33}$$

With this definition, the phase function is normalized,

$$\int_{4\pi} P(\cos\theta) \frac{d\omega}{4\pi} = 1. \tag{3.34}$$

3.3.3 Geometric optics and anomalous diffraction

We now turn to approximate theories for large particles that lean heavily upon ideas about geometric optics and ray theory. For this discussion it is convenient to change the presentation. First, we redefine all coefficients to be per sphere rather than per molecule. This requires no more than a change in definition of n in equation (3.1), from the number of molecules per unit volume to the number of spheres per unit volume.

Second, we normalize with the cross section of the sphere, and discuss dimensionless *scattering*, *absorption*, and *extinction efficiencies*,

$$Q_s = \frac{s}{\pi r^2}, \tag{3.35}$$

$$Q_a = \frac{k}{\pi r^2}, \tag{3.36}$$

$$Q_e = Q_a + Q_s. \tag{3.37}$$

In §3.1 we identified the extinction coefficient with the cross section for the interaction between photon and particle. When geometric optics is strictly valid, and the wave nature of light is neglected, the cross section for a sphere is equal to the shadow area, πr^2. Hence, for a refracting but nonabsorbing sphere, that disturbs all rays incident upon it, $Q_s = 1$, $Q_a = 0$, $Q_e = 1$; while for a black sphere, $Q_s = 0$, $Q_a = 1$, $Q_e = 1$.

But we cannot neglect light passing close to the sphere, and it is necessary to supplement geometric optics by including *diffraction* as an independent phenomenon. While this cannot be justified analytically, the usefulness of the procedure is attested to by the success of elementary physical optics.

There is a well-known result of wave optics that, for a circular aperture, the Poynting vector at scattering angle θ is,

$$N(\theta) \propto \left[\frac{J_1(x \sin \theta)}{x \sin \theta}\right]^2, \qquad (3.38)$$

where J_1 is a Bessel function of integral order. This quantity is shown by the broken line in Figure 3.13. Most of the light is concentrated into a strong forward peak, sometimes called a *diffraction peak*. For angles away from the forward direction there are a series of weak maxima and minima that represent a series of *diffraction rings* about the incident direction. The first minimum is close to $x \sin \theta = 4$ so that when $x = 40$ ($r = 3.2$ μm at $\lambda = 500$ nm), for example, most diffracted light is within $\sim 6°$ of the incident direction. Although this scattering angle is small, the light has, nevertheless, been deviated, and Mie theory will include all diffracted light in the calculation of the scattering coefficient.

A theorem by Babinet states that the total energy diffracted by a sphere is exactly equal to the amount intercepted by the sphere, so that, to take account of diffraction in the limit of geometric optics, we may simply add unity to both scattering and extinction efficiencies. For both transparent and black spheres, the extinction efficiency is equal to two in the limit of very large particles.

According to the Kirchoff-Fresnel treatment of wave propagation, a new wave front may be constructed from the interference of secondary waves from a previous wave front. With nothing to introduce phase lags, a plane wave will follow from a plane wave. A translucent sphere interferes with this picture. We are interested in all changes from the original plane wave front, because this is what the definition (3.1) refers to.

Van de Hulst's theory of anomalous diffraction simplifies this problem by considering a sphere with both $x \gg 1$, and $|\widetilde{m} - 1| \ll 1$, but with a finite size for the product of the two. The second condition means that rays striking the sphere are neither reflected nor refracted, because the refractive indices inside and outside the sphere are similar. The finite product allows,

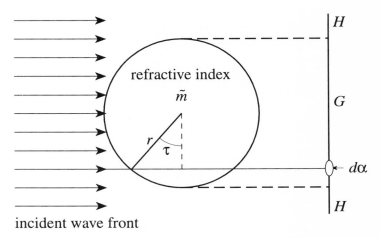

Figure 3.11 Geometry of anomalous diffraction.

at the same time, a finite value to the angular phase lag, equal to

$$\rho = 2\pi \frac{2r|\widetilde{m} - 1|}{\lambda} = 2x|\widetilde{m} - 1|$$

for rays crossing a diameter of the sphere, see Figure 3.11.

By eliminating reflection we have also eliminated any possibility of polarization in this formulation. Unlike small-particle scattering, anomalous diffraction is, in principle, not polarized.

Let $E_{\theta=0}(\widetilde{m})$ be the electric field in the direction $\theta = 0$ at the final wave front; the corresponding quantity for the incident wave front is taken to be unity. We divide the final wave front into that which is within the geometric shadow area, G, and that which lies outside, H. The area of G is πr^2. The total final field is,

$$\int_{G+H} E_{\theta=0}(\widetilde{m}) \, d\alpha = \int_{G} E_{\theta=0}(\widetilde{m}) \, d\alpha + \int_{H} E_{\theta=0} \, d\alpha \,. \tag{3.39}$$

The only vector property of the electric field that we shall take account of is the phase change of a ray crossing the sphere.

Now consider equation (3.39) for the limiting case $\widetilde{m} \equiv 1$, when there are no longer any phase changes caused by the material in the sphere. The left side must become $\int_{G+H} d\alpha$ because the final wave front must be the same as the incident wave front, $E_{\theta=0} = 1$. However, the terms on the right side are not *individually* equal to $\int_{G} d\alpha$ and $\int_{H} d\alpha$, respectively. The reason for this is that the second term cannot be a function of $\widetilde{m}$ since none of the rays involved pass through the sphere, and cannot experience any change associated with the material in the sphere. We know that radiation in the region H is affected by diffraction if G is an obstacle. The electric field in H is disturbed and the second term on the right side of (3.39) must differ from $\int_H d\alpha$. The foregoing discussion shows that the difference is the same even if the obstacle is completely transparent. The difference arises from division of the wavefront into sections G and H regardless of optical properties of G.

We now evaluate the change in $E_{\theta=0}$ when a dielectric sphere is added to a previously empty space. This will give us the attenuation of a beam falling on the sphere, but continuing in the direction $\theta = 0$, which will enable us to calculate the extinction coefficient. Electrical fields will also appear at other angles (the scattered field), but these are not our first concern. To evaluate the difference in fields we subtract equation (3.39) with $\widetilde{m} \equiv 1$. From (3.39), the integrals over H cancel, and we find,

$$\int_{G+H} [E_{\theta=0}(\widetilde{m}) - 1]\, d\alpha \equiv \int_{G+H} \delta E\, d\alpha,$$
$$= \int_G [E_{\theta=0}(\widetilde{m}) - E_{\theta=0}(1)]\, d\alpha. \quad (3.40)$$

If we take account of the phase shift along undeviated rays, we may write,

$$E_{\theta=0}(\widetilde{m}) = E_{\theta=0}(1) e^{i\rho \sin \tau}, \quad (3.41)$$

where the phase shift parameter is,

$$\rho = 2x(\widetilde{m} - 1), \quad (3.42)$$

and τ is the angle illustrated in Figure 3.11. Hence,

$$\int_{G+H} \frac{\delta E}{E}\, d\alpha = \int_G \left(e^{i\rho \sin \tau} - 1\right) d\alpha. \quad (3.43)$$

With $d\alpha = 2\pi r^2 \cos \tau \sin \tau\, d\tau$, the right side of (3.43) may be integrated,

$$\int_{G+H} \frac{\delta E}{E}\, d\alpha = -2\pi r^2 K(i\rho), \quad (3.44)$$

where, after some manipulation, it may be shown that,

$$K(w) = \frac{1}{2} + \frac{e^{-w}}{w} + \frac{e^{-w} - 1}{w^2}.$$

Finally, we must interpret (3.44) in terms of the extinction coefficient. Since $I \propto |E|^2$, it is a straightforward exercise in complex numbers to show

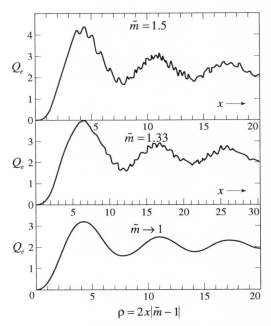

Figure 3.12 Extinction by large dielectric spheres. The top two panels are Mie calculations. The bottom panel is anomalous diffraction from equation (3.46). For dielectric spheres $Q_e = Q_s$.

that $\frac{\delta I}{I} = 2\Re\left(\frac{\delta E}{E}\right)$, where $\Re$ represents the real part. If we now apply (3.1) to a single scatterer, as we did in the derivation of equation (3.28), we find,

$$-\frac{\delta I}{I} = -2\Re(\frac{\delta E}{E}) = e \ . \tag{3.45}$$

If we restrict ourselves to the case of real ρ, that is, to scattering, Q_e may be shown to be,

$$Q_e = Q_s = \frac{e}{\pi r^2} = 2 - \frac{4}{\rho}\sin\rho + \frac{4}{\rho^2}(1 - \cos\rho) \ . \tag{3.46}$$

Although we shall give no details, the above derivation may be extended to complex refractive indices, that is, to absorbing spheres and to off-axis angles, in a straightforward manner. Details may be found in van de Hulst (1957). We shall now look at some of the results that may be calculated from van de Hulst's theory of anomalous diffraction.

Figure 3.12 shows a comparison between calculations based on equation (3.46) and two Mie theory (i.e., exact) calculations for spheres with real refractive indices (also called *dielectric spheres*). Although it is hard to see, exact and approximate calculations do not agree well for $x \ll 1$,

ABSORPTION AND SCATTERING

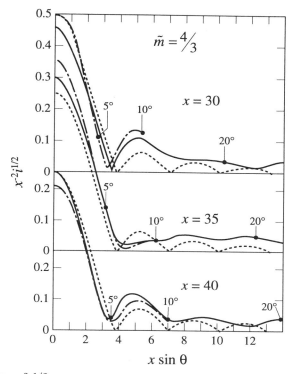

Figure 3.13 $x^{-2}i^{1/2}$, as a function of $x\sin\theta$. i stands for either i_1 or i_2. The refractive index is $\frac{4}{3}$. The heavy solid lines with points are exact calculations. The dash-dot lines are for anomalous diffraction. The broken lines are Fraunhofer diffraction patterns. Vertical tags indicate angles.

but this is to be expected because the Kirchoff-Fresnel treatment is unsatisfactory in this limit. For larger values of x, anomalous diffraction (in the bottom panel) captures the essence of the behavior of the scattering efficiency, even for refractive indices substantially different from unity. It fails to reproduce fine structure but, because of the mixture of particle sizes occurring in natural aerosols, this fine structure is not important in atmospheric applications.

For $x > 1$, all curves show a wave-like structure that is caused by interference between transmitted and diffracted light. The separation between successive maxima or minima is $\Delta\rho \sim 2\pi$ (ρ is the phase lag expressed in angular units). For $x > 5$ the variation of Q_e with x is not large. Since $x \propto \lambda$, this implies that color effects will not be large; the scattering is "whiter" than for Rayleigh scattering. For $x \gg 1$ the geometric optics limit is reached: $Q_e \to 2$ as $x \to \infty$, and the scattering shows no color effects. For a dielectric sphere there is no absorption so that $Q_s \to 2$.

The angular distribution of scattering, as predicted by anomalous diffraction, is compared to exact calculations and to Fraunhofer diffraction in Figure 3.13. The agreement between anomalous diffraction and exact calculations is fair but, most importantly, the strong forward scattering peak that we associated with diffraction is clearly demonstrated in all of the calculations.

3.3.4 Large- and small-particle extinction

Scattering and absorption by suspended particles, from molecules to cloud droplets, is one of the most critical of climate elements, and the least well understood. Scattering centers in the atmosphere span a range of sizes from molecules (4×10^{-10} m), to the smallest condensation nuclei ($\sim 10^{-8}$ m, see Chapter 8), to cloud droplets ($\sim 10^{-4}$ m), and relevant wavelengths span a range from the near ultraviolet (2×10^{-7} m) to the far infrared (2×10^{-4} m). The range of values of the size parameter, x (eight to nine orders of magnitude), the variable compositions, and the variable shapes of the suspended particles are such that no "complete" theory of atmospheric extinction is likely ever to be achieved. We are obliged to think in terms of generalities, and the two limiting cases of Rayleigh scattering and anomalous diffraction are probably the most important. We may compare them qualitatively as follows:

Polarization. Small-particle scattering shows strong polarization at certain scattering angles. A satisfactory theory of the light from the daytime sky is impossible without taking polarization into account. Anomalous diffraction is not polarized; large-particle scattering can, in general, show polarization, but it is of minor importance.

Color. Single-scattering from small particles is blue, while transmitted light is red. The brilliant colors of twilight are partly caused by molecular scattering. For very large particles, scattering is independent of the size parameter and, for incident sunlight, it is white.

Direction. No natural particles give exactly isotropic scattering, but small particles come close, and departures from isotropy are of small importance, except where polarization is concerned. For large particles, on the other hand, approximately half of the scattering goes into a strongly forward diffraction peak. We have seen that, for large droplets, the angular size of this peak may be very small. For some purposes small-angle scattering may even be regarded as not belonging to the extinction process at all, and may be neglected. To illustrate this point consider the bright *aureole* around the solar disc when seen through cloud or mist, which is a manifestation of forward scattering by large particles. We would not anticipate any serious error for thermal calculations if the aureole were taken to be part of the undeviated solar beam.

ABSORPTION AND SCATTERING

Table 3.2 Some scattering regimes

Regime	Conditions	Q_s	ns
Rayleigh	$x \ll 1$	$\frac{8}{3}x^4 \left|\frac{\widetilde{m}^2-1}{\widetilde{m}^2+2}\right|^2$	$\propto x^3$
Rayleigh-Gans	$x \gg 1,\ \rho \ll 1$	$2x^2|\widetilde{m}-1|$	$\propto x$
Anomalous diffraction	$x \gg 1,\ \rho$ finite	2	$\propto x^{-1}$

Rayleigh scattering is given in the form appropriate for a single particle; see the text.

Finally, we may ask the question: For a given amount of matter, for example, a given amount of water vapor, what size of particle will lead to most scattering? Table 3.2 shows the scattering coefficient per unit volume, ns, for spheres of refractive index, $\widetilde{m}$, when the total volume of condensed material, $n\frac{4}{3}\pi r^3$, is held constant, so that $n \propto x^{-3}$. Three asymptotic regimes are considered; the Rayleigh expression is obtained from equation (3.30), using the polarizability of a single dielectric sphere, $p = \frac{\widetilde{m}^2-1}{\widetilde{m}^2+2}r^3$, so that all three regimes have a common point of reference. The Rayleigh-Gans expression has not been discussed here; it is for large particles that exhibit negligible phase shifts; it is of no practical interest but helps to define the trend of scattering as a function of particle size. The anomalous diffraction is only given in its limiting form for large x.

According to the fourth column in Table 3.2, the scattering power per unit volume, holding the amount of material fixed, grows rapidly with particle size for small particles, continues to grow under some conditions for large particles but, eventually, decreases, as large drops appear. Maximum scattering per unit volume will be for drops whose diameters are on the order of magnitude of the wavelength of the scattered radiation ($x \sim 1$).

3.4 Reading

General books on the subject of atmospheric radiation are

Goody, R.M., and Yung, Y.L., 1989, *Atmospheric radiation: Theoretical basis,* 2nd ed. New York: Oxford University Press.

Liou, K.N., 1980, *An introduction to atmospheric radiation.* New York: Academic Press.

It is difficult to find a readable introduction to molecular spectroscopy that is neither out of date nor too specialized. One of the most useful accounts is

Townes, C.H., and Schawlow, A.L., 1982, *Microwave spectroscopy*. New York: Dover Publications.

This book is not as restricted in scope as the title might suggest.

Two valuable monographs on scattering are

Van de Hulst, H.C., 1957, *Light scattering by small particles*. New York: Wiley.

Bohren, C.F., and Huffman, D.R., 1983, *Absorption and scattering of light by small particles*. New York: Wiley.

3.5 Problems

Asterisks* and double asterisks** indicate higher degrees of difficulty.

3.1 **Equivalent widths.** This discussion refers to absorption by a single, isolated absorption line (see Figure 3.14). The absorption area, as indicated in the figure, is known as the *equivalent width*, W. It is proportional to the energy absorbed by an isolated line. For this problem the absorption path (length l, number density of absorbers, n) is homogeneous, that is, k_ν is constant.

(i) Write a formal expression for the equivalent width of an isolated, Michelson-Lorentz line of strength, S, in terms of the dimensionless variables:

$$u = \frac{Snl}{2\pi\alpha_L}; \quad x = \frac{\nu - \nu_0}{\alpha_L} - A.$$

For all vibration-rotation bands, $\frac{\nu_0}{\alpha_L} \gg 1$.

What is the physical significance of the quantity $2u$?

(ii) The expression in (i) cannot be integrated in terms of known functions, but it can be integrated in the *strong line* and *weak line limits*, when the absorption in the line center is very strong or very weak, respectively. Obtain expressions for the equivalent width in these to limits in terms of the dimensional variables.

$$\int_0^\infty \left\{1 - \exp\left(\frac{1}{y^2}\right)\right\} dy = \pi^{1/2}.$$

3.2 **Detection of deuterium.*** The outer atmosphere of Saturn consists of atomic hydrogen with a very small admixture of atomic deuterium

ABSORPTION AND SCATTERING

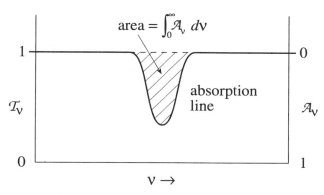

Figure 3.14 Equivalent width. $\mathcal{T}_\nu$ is the transmission, and $\mathcal{A}_\nu$ is the absorption, see §3.1.

at very low pressures, and at a temperature of 1000 K. D and H have resonance lines at 121.533 nm and 121.566 nm, respectively, with effectively the same line strengths. The light from an occulted star reaches a spacecraft through such a long path that the deuterium line would be measurable with 1% or 2% absorption at the line center if not swamped by the wing absorption of the hydrogen line.

(i) What are the Doppler line widths (in nanometers) of the D and H lines?

(ii) At what D/H ratio will there be equal absorption by both gases at the center of the deuterium line?

(iii) The absorption is measured with a spectrometer that smooths the absorption out evenly over a wavelength range of 0.01 nm (the *spectral resolution* of the instrument). Make an order-of-magnitude estimate of the D/H ratio for equal absorption under this circumstance.

3.3 The random band model.** Band models refer to formal arrays of lines for which the mean absorption or transmission over many lines may be obtained analytically. The use of band models is now largely replaced by the correlated-k technique described in §4.3.1. One important class of model is the random model, which we discuss in this problem.

Figure 3.15 shows a spectral interval of width $N\delta$ containing N identical absorption lines, placed at random within the interval, with a mean spacing δ. The equivalent width of each line is the same, W. For convenience, the zero of frequency has been moved to the band center.

(i) What is the mean absorption, $\overline{A}$, averaged over the entire region, if the spectral lines do not overlap?

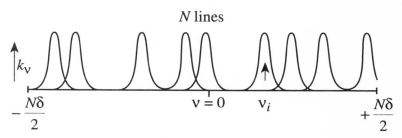

Figure 3.15 A random band.

(ii) If lines have centers at ν_i, $i = 1....N$, what is the transmission for path length l at $\nu = 0$? If the lines are randomly placed in the interval, the probability that a line lies between ν_i and $\nu_i + d\nu_i$ is $\frac{d\nu_i}{\delta}$. Obtain an expression for the mean transmission of an array averaged over all random realizations of the array, and allow the number of lines in the band to go to infinity. This is one example of a random band.

$$\left(1 - \frac{x}{N}\right)^N \to \exp(-x) \quad \text{as} \quad N \to \infty \ .$$

(iii) Under what condition it is possible to neglect line overlap for a random band?

3.4 Once in a blue moon. This expression has a basis in fact. On very rare occasions the moon has been observed to have a blue coloration. The last occasion known to the writer was during the 1950s, when blue moons were observed in Europe following very intense forest fires in Canada.

According to the discussion in §8.3.1, the commonest type of atmospheric aerosol is the Aitken nucleus, with a radius less than 0.1 μm. A major source of Aitken nuclei is the condensation of gaseous combustion products. Water aerosols, on the other hand, have radii $\sim$ 10 μm, see Figure 8.1.

(i) Can you find an explanation for the blue moon in terms of the universal properties of particulate scattering? Consider the data for $\widetilde{m} = 1.33$ in Figure 3.12.

(ii) What physical properties must you attribute to the scattering particles? Why do you think that they are rare?

3.5 Measurement of extinction. A "parallel" beam of radiation, of wavelength 0.5 μm, passes through a scattering cell containing nonab-

ABSORPTION AND SCATTERING

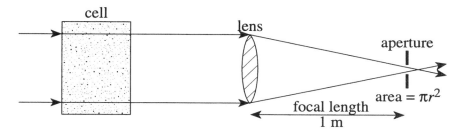

Figure 3.16 An experimental arrangement for extinction measurement.

Table 3.3 Extinction data

λ nm	$k(O_3)$, m^2	s_R, m^2	τ
300	3.16×10^{-23}	5.59×10^{-30}	4.083
400	negligible	1.67×10^{-30}	0.515
500	1.15×10^{-25}	6.66×10^{-31}	0.253
600	2.47×10^{-25}	2.29×10^{-31}	0.139
700	9.75×10^{-26}	1.70×10^{-31}	0.0955
800	2.05×10^{-26}	9.88×10^{-32}	0.0620

sorbing droplets, of a single size. The light is focused by an optical system of focal length 1 m onto a small circular aperture of radius, r. The source is extended enough to fill the aperture. See Figure 3.16.

The extinction coefficient is measured by recording the energy flux through the aperture, with and without the scattering cell, and the use of equation (3.2). The measured scattering coefficient is found to depend upon the radius of the aperture. For $r \ll 2$ mm the scattering coefficient measured is approximately twice that measured when $r \gg 2$ mm. Explain this observation. What is the approximate size of the scattering particles?

3.6 **Atmospheric extinction.*** Table 3.3 shows ozone absorption coefficients, $k(O_3)$, and Rayleigh scattering coefficients, s_R, for wavelengths of 300 to 800 nm. In addition to ozone and air molecules, extinction is also caused by an aerosol having an extinction coefficient per particle of 3×10^{-10} m^2 at 400 nm, and a variation with wavelength $e_\lambda \propto \lambda^{-2}$. Also shown in Table 3.3 is the optical depth of the entire atmosphere ($\tau = \int_0^\infty ne\,dz$), obtained from direct measurements. How many air molecules, ozone molecules, and particles are there per unit area in the whole atmosphere?

CHAPTER 4

RADIATIVE TRANSFER

In order to calculate radiation fields we need to describe in mathematical terms how radiation is both created and destroyed. The loss of radiation is described by Lambert's law and the processes discussed in Chapter 3. The creation of photons, or emission, is described formally in terms of the radiance emitted by unit optical path, the source function. Given the source function and extinction coefficients, we may construct a differential equation, Schwartzschild's equation of transfer, or an algorithm suitable for finite differences in a stratified atmosphere; the latter requires a logical extension of Lambert's law, known as the interaction principle.

The source function depends upon the nature of the interactions between photons and molecules. The two principal classes of interaction are scattering and absorption. Each process has its own source function, and the complete source function may be stated as a sum of source functions for these two processes taken independently.

Absorption results in the transfer of energy from the radiation field to the thermal reservoir. The associated emission process is thermal emission, for which energy goes from the thermal reservoir into the radiation field. If the molecules are in local thermodynamic equilibrium, the source function is the Planck function, a known function of frequency and temperature only.

The scattering source function is the result of photons being redirected from incident photons by the scattering process. Since photons can be incident from any direction, the scattering source function involves an integral over all directions of incidence. This feature of the scattering process greatly increases the complexity of radiation calculations.

The remainder of Chapter 4 discusses commonly used methods for solving the transfer problem. For a thermal equilibrium source function, the equation of transfer can be solved in terms of a simple quadrature. This is the method of choice for most weather and climate models. If scattering is to be explicitly included in the calculation, it is more convenient to use the interaction principle. One way to do so is the doubling-and-adding method.

This method involves fairly complicated matrix algebra, but modern computers have the rapid algorithms that can handle them with facility.

Despite the existence of "exact" solutions, the need to approximate is of paramount importance. The number of monochromatic calculations that are needed for molecular line spectra is so great that they can only be contemplated for rare occasions. The correlated-k method solves this problem by approximating frequency integrals over line spectra by sums over a few finite intervals of a new independent variable.

Approximations to the transfer equation may be based upon the method of moments. The resulting differential equations give a good qualitative account of the physics of radiative transfer, and are relatively easy to solve, providing the investigator with a back-of-the-envelope tool. The simplest approximation of all is Newtonian cooling, an explicit statement of radiative cooling that can be remarkably satisfactory, and is valuable for analytical studies in dynamical meteorology.

4.1 Fundamental ideas

4.1.1 Radiance and the equation of transfer

The quantity used to describe diffuse fields of radiation, whether thermal or scattered, is the *energy intensity* or *radiance*. Consider the "telescope" illustrated in Figure 4.1. It has an objective of area, A, centered on the point, P, and an aperture of area, α, at the focal plane of the objective, whose focal length is f. The telescope accepts radiation from a small solid angle, $d\omega_l$, surrounding the l-direction, defined by the aperture,

$$d\omega_l = \frac{\alpha}{f^2}. \tag{4.1}$$

Behind the aperture is a filter that limits frequencies to between ν and $\nu + d\nu$, and a detector to measure the rate, E, at which energy passes through the filter. E must be proportional to A, to $d\nu$, to α and, therefore, to $d\omega_l$, provided that each is small. The constant of proportionality is, by definition, the energy intensity or radiance,

$$I_\nu(P, \vec{l}) = \frac{E}{A d\nu d\omega_l}. \tag{4.2}$$

Lambert's law, equation (3.1), is the fundamental law for the *loss* of radiance as a result of interactions between radiation and matter. Thermal emission and scattering, on the other hand, are both processes that *add* to the radiance. Both thermal emission and scattering add radiance proportionally to the number of radiating molecules per unit area in the path,

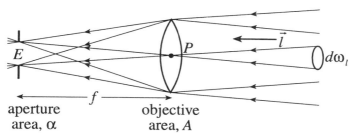

Figure 4.1 Definition of radiance.

$n\,dl$. As a matter of formal definition, we introduce the *source function*, J_ν, such that the emission adds to the radiance the amount,

$$dI_\nu(P,\vec{l}) = +J_\nu(P,\vec{l})e_\nu n\,dl\;. \tag{4.3}$$

Note the positive sign in (4.3), indicating emission. The inclusion of the extinction coefficient, e_ν, in the definition is largely formal.

All interactions between radiation and matter result in either extinction or emission. We may, therefore, add equations (3.1) and (4.3) to obtain a differential equation for the total change of radiance resulting from all interactions between radiation and matter. This is *Schwartzschild's equation of transfer*,

$$\frac{1}{ne_\nu}\frac{dI_\nu(P,\vec{l})}{dl} = J_\nu(P,\vec{l}) - I_\nu(P,\vec{l})\;. \tag{4.4}$$

Schwartzschild's equation was accepted as the best statement of Lambert's law and the first law of thermodynamics before numerical methods came into common use for atmospheric calculations. Numerical models of the atmosphere employ finite slabs rather than continuous variables, and a differential equation is not an ideal starting point for such methods. Consequently, for numerical models, these two fundamental laws have been reformulated for stratified slabs under the name of the *interaction principle*. We shall discuss these matters in §4.2.2.

4.1.2 Radiation flux and heating rate

From the thermodynamic standpoint, the most important property of the radiation field is the flux of radiant energy, $\vec{F}(\text{rad})$, integrated over all frequencies. For the remainder of this chapter, we shall use the abbreviated notation $\vec{F}$ for the radiant flux, and $\vec{F}_T$ and $\vec{F}_S$ to mean the thermal and solar fluxes, respectively. First, we need to derive the monochromatic flux, $\vec{F}_\nu$.

The component of the flux in the d-direction is the total energy from all directions flowing across a unit area whose surface normal is in the d-direction; see, for example, the geometry in Figure 4.3. From the definition of radiance the d-component of $\vec{F}_\nu$ is,

$$F_{\nu,d} = \int_{4\pi} I_\nu(\vec{l})\xi_{d,l}\, d\omega_l ,\qquad(4.5)$$

where $\xi_{d,l}$ is the d-component of a unit vector in the l-direction, that is, the cosine of the angle between the two vectors $\vec{l}$ and $\vec{d}$. From equations (4.4) and (4.5), the flux divergence is,

$$\begin{aligned}
\nabla \cdot \vec{F} = \int_0^\infty \nabla \cdot \vec{F}_\nu\, d\nu &= \int_0^\infty \frac{\partial F_{\nu,i}}{\partial x_i}\, d\nu,\\
&= \int_0^\infty d\nu \int_{4\pi} \frac{\partial I_\nu}{\partial x_i}\xi_i\, d\omega_l,\\
&= \int_0^\infty d\nu \int_{4\pi} \frac{dI_\nu}{dl}\, d\omega_l,\\
&= \int_0^\infty d\nu \int_{4\pi} ne_\nu(J_\nu - I_\nu)\, d\omega_l .\qquad(4.6)
\end{aligned}$$

The sum rule for repeated indices has been used.

It is possible, in principle, to have an atmospheric state in which there is no heat transport by motions or by any means other than by radiation. This is a state of *radiative equilibrium*. The radiative heating rate must then be zero. According to equation (2.1), this means that the flux divergence is zero. From (4.6) the most general requirement for radiative equilibrium is,

$$\int_0^\infty e_\nu\, d\nu \int_{4\pi} J_\nu\, d\omega = \int_0^\infty e_\nu\, d\nu \int_{4\pi} I_\nu\, d\omega .\qquad(4.7)$$

In §4.1.5 we shall discuss a version of (4.7) without the frequency integration. This defines *monochromatic radiative equilibrium*. We shall find that the scattering source function always satisfies this condition.

4.1.3 The thermal source function

The physical content of the equation of radiative transfer is contained in the extinction coefficient and the source function. The former was discussed in Chapter 3; we now consider the source functions for absorption and scattering. If the two processes occur simultaneously, as will normally be the case, the radiances will add. Since $e_\nu = k_\nu$ for absorption, and $e_\nu = s_\nu$ for scattering, it follows that,

$$e_\nu J_\nu = s_\nu J_\nu(scattering) + k_\nu J_\nu(thermal)$$

or,
$$J_\nu = a_\nu J_\nu(scattering) + (1 - a_\nu) J_\nu(thermal) , \qquad (4.8)$$

where a_ν is the single scattering albedo, for a single frequency, see equation (3.5). The distinction between $J_\nu(scattering)$ and $J_\nu(absorption)$ is formal, since neither process occurs alone; the former may be calculated by assuming that k_ν is zero, and the latter by assuming that s_ν is zero.

Kirchoff was the first person to apply the laws of thermodynamics systematically to the radiation field inside a constant-temperature enclosure or cavity. This is a state of complete thermodynamic equilibrium. Kirchoff showed that the radiance is isotropic and homogeneous, that it depends only upon the frequency of the radiation and the temperature,[1] and that the radiance and the source function are the same. The symbol $B_\nu(T)$ is used for this equilibrium source function. The radiation field has several names: *equilibrium radiation*, *cavity radiation*, and *black-body radiation*. The last name follows from Kirchoff's demonstration that the radiation emitted from a black surface is the same as cavity radiation at the same temperature.

The form of $B_\nu(T)$ was established by Planck (the *Planck function*),

$$B_\nu(T) = \frac{2h\nu^3}{c^2} \frac{1}{\exp\left(\frac{h\nu}{kT}\right) - 1} , \qquad (4.9)$$

where **h**, **k**, and **c** are Planck's constant, Boltzmann's constant, and the speed of light, respectively.

The Planck function is plotted in three different ways in Figure 4.2. It has a single maximum at a wavelength that is given by *Wien's displacement law*,

$$T\lambda_{max} = 2.8978 \times 10^{-3} \text{ m K} . \qquad (4.10)$$

It behaves differently in the two wings. As $\lambda \to \infty$, in the infrared spectrum,

$$B_\lambda \to \frac{2kTc}{\lambda^4} . \qquad (4.11)$$

This is the *Rayleigh-Jeans radiation law*. As $\lambda \to 0$, in the ultraviolet spectrum,

$$B_\lambda = \frac{2hc^2}{\lambda^5} \exp\left(-\frac{hc}{\lambda kT}\right) , \qquad (4.12)$$

which is the *Wien radiation law*. The most important distinction between these two regimes is the rapidity of changes involving the temperature and the frequency in the Wien radiation law, because of the exponential term.

---

[1] It is also a function of the refractive index of the medium but, for air, the refractive index is essentially unity.

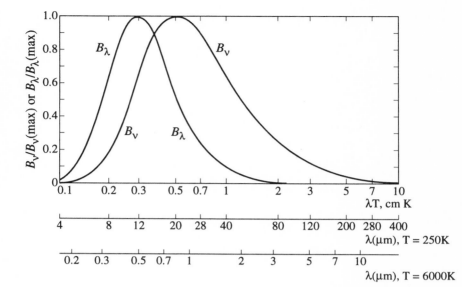

Figure 4.2 The Planck function. B_ν is from equation (4.9). $B_\lambda = B_\nu |\frac{d\nu}{d\lambda}| = B_\nu \frac{c}{\lambda^2}$. The maxima are:

$$\pi B_\lambda(\text{max}) = 1.2868 \times 10^{-5}\, T^5 \text{ W m}^{-2}\text{ m}^{-1},$$
$$\pi B(\text{max}) = 5.9566 \times 10^{-19}\, T^3 \text{ W m}^{-2}\text{ Hz}^{-1}.$$

Equation (4.9) may be integrated over frequency to give the *Stefan-Boltzmann law*,

$$B(T) = \int_0^\infty B_\nu(T)\, d\nu = \frac{\sigma}{\pi} T^4, \qquad (4.13)$$

where

$$\sigma = \frac{2\pi^5}{15}\frac{k^4}{c^2 h^3},$$

is the Stefan-Boltzmann constant. The flux of radiation from unit area of a black surface is the integral (4.5) taken over one hemisphere (see Figure 4.3),

$$F^+ = \int_0^\infty d\nu \int_0^1 2\pi B_\nu(T) \xi\, d\xi = \pi B(T) = \sigma T^4. \qquad (4.14)$$

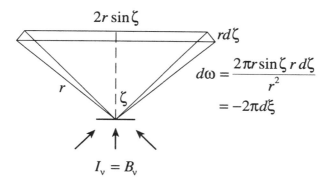

Figure 4.3 Hemispheric radiation from a black surface. When surface is horizontal, ζ is the *zenith angle*.

4.1.4 The scattering source function

Scattering is the result of reorientation of radiant energy from an incident direction $\vec{d}$ to a scattered direction $\vec{l}$. In the absence of absorption, all of the radiation lost must reappear in the scattered radiation field.

Let an incident beam of ν-radiation be confined to a small solid angle $d\omega_d$ in the d-direction. The energy flow *per unit area* per second is $I_\nu(\vec{d})\,d\omega_d$. From equation (3.1), the loss of incident energy *per unit volume* per second is $s_\nu I_\nu(\vec{d})n\,d\omega_d$. From the definition of the phase function,[2] equation (3.32), the fraction of the radiation scattered into a solid angle $d\omega_l$ in the l-direction is $P(\theta_{d,l})\frac{d\omega_l}{4\pi}$. $\theta_{d,l}$ is the angle between the d- and the l-directions. If we integrate over all incident directions, we find for the total radiation scattered in the l-direction, $ns_\nu\,d\omega_l \int_{4\pi} I_\nu(\vec{d})P(\theta_{d,l})\frac{d\omega_d}{4\pi}$.

By similar reasoning from equation (4.3), the total scattered radiation per unit volume that flows into $d\omega_l$ may be written in the alternate form, $J_\nu(\vec{l})s_\nu n\,d\omega_l$. Equating the two statements gives,

$$J_\nu(\vec{l})(scattering) = \int_{4\pi} I_\nu(\vec{d})P(\theta_{l,d})\frac{d\omega_d}{4\pi}. \qquad (4.15)$$

If we substitute (4.15) into (4.7), and use the normalization condition for the phase function, equation (3.34), we find that the scattering source function leads to a state of monochromatic radiative equilibrium. This must be so because scattering allows no energy to enter the thermal reservoir, but it is reassuring that our ideas are consistent.

It is convenient to separate out that part of the source function for scat-

---

[2] We make no attempt to give the formalism suitable for polarization studies. If this were required, the radiance would have to be replaced by a vector of four radiances called *Stokes parameters*, and the phase function by a 4×4 matrix.

tering that is the result of primary scattering from the direct solar beam. Intuitively, there is a qualitative difference between the quasi-parallel, "direct" solar radiation, and the diffuse, scattered, solar radiation. Both contribute to the radiance on the right-hand side of equation (4.15), but the operation of the integral upon the two components is different, and they are usually separated.

The total radiance is the sum of the radiance in the direct beam, $I_{\nu,\odot}$, and that of the diffuse radiation, $I_{\nu,d}$. The source function may be separated into two terms, one with each radiance substituted into (4.15). If $I_{\nu,\odot}$ is substituted, (4.15) yields a source function, $J_{\nu,p}$, the *source function for primary scattering*. As the name implies, this source function describes the scattered radiation that has been scattered once only. The *source function for multiple scattering*, $J_{\nu,m}$, is obtained by substituting $I_{\nu,d}$ in equation (4.15). Both $J_{\nu,m}$ and $J_{\nu,p}$ contribute to the scattered radiance and the scattering source function is the sum of the two.

$I_{\nu,\odot}$ comes from the solar disc, and is zero except within a small solid angle, $d\omega_\odot$, in the direction of the solar beam, $\vec{d}_\odot$. Substitution in equation (4.15) gives,

$$J_{\nu,p}(P,\vec{l}) = I_{\nu,\odot} P(\theta_{d_\odot,l}) \frac{d\omega_\odot}{4\pi} . \qquad (4.16)$$

We now define the *solar irradiance*, f_ν, to be the flux of radiation from the sun, in the direction of the solar beam. This is the quantity that is usually given in the literature to describe the output of radiation from the sun. It varies seasonally with the inverse square of the earth-sun distance. Reduced to the mean earth-sun distance, it is sometimes referred to as the *solar constant*.[3] In terms of the radiance of the sun, the definition of flux, (4.5), gives,

$$f_\nu = \int_{4\pi} I_{\nu,\odot}(P,\vec{l}) \xi_{d_\odot} \, d\omega_l = I_{\nu,\odot} \, d\omega_\odot , \qquad (4.17)$$

and the source function for primary scattering is then,

$$J_{\nu,p}(P,\vec{l}) = f_\nu \frac{P(\theta_{d_\odot,l})}{4\pi} . \qquad (4.18)$$

We assume the solar beam to be very narrow, and the amount of diffuse radiation, thermal or scattered, that falls within $d\omega_\odot$ may be neglected. This means that a transfer equation for f_ν contains attenuation but no emission. This takes us back to the extinction process discussed in §3.1. If we write $dl' = \frac{dz'}{\xi_\odot}$, as illustrated in Figure 4.5, we may integrate from z to ∞ and, from equation (3.2), the irradiance at height z is,

$$f_\nu(z) = f_{\nu,0} \exp\left(\int_z^\infty e_\nu(z') n(z') \frac{dz'}{\xi_\odot} \right) . \qquad (4.19)$$

$f_{\nu,0}$ is the irradiance in space.

---

[3]Imprecisely, because the sun is not a constant source, see Appendix D.

RADIATIVE TRANSFER

4.1.5 The complete source function

If we accept the distinction between scattering and thermal source functions (but see the next section for a more flexible view), we may now write a complete source function from (4.8), (4.15), and (4.18). For the direct solar beam, (4.19) provides a complete solution, and it is not necessary to include the direct solar radiance in radiative transfer calculations. Hence, from this point we understand the radiance I_ν to be the diffuse radiation $I_{\nu,d}$. With this understanding,

$$J_\nu(P, \vec{l}) = (1 - a_\nu)B_\nu + a_\nu \int_{4\pi} I_\nu(P, \vec{d}) P(\theta_{d,l}) \frac{d\omega_d}{4\pi} + a_\nu f_\nu \frac{P(\theta_{d_\odot,l})}{4\pi} \ . \quad (4.20)$$

4.1.6 Nonequilibrium source functions

If the emission of matter were the same outside a constant-temperature enclosure as it is inside the enclosure, the source function outside the enclosure would be the thermal equilibrium source function, $B_\nu(T)$. However, the ambient radiation field is different in the two cases and, according to Einstein's theory of *induced emission*, radiation is induced by incident photons. The thermal source function should, therefore, depend upon the incident radiation and should not be the same inside and outside the constant-temperature cavity. Nevertheless, the equilibrium source function, $B_\nu(T)$, is always used in meteorological calculations, and we now examine the reasons why.

Some of the considerations relevant to this question were raised in our discussion of local thermodynamic equilibrium in §1.4.2 and §1.4.3, and of photolysis in §1.5.1. Let us disregard chemical reactions. Then processes 1 and 2 in Table 1.4 compete for the fate of an incident photon. Process 1 leads to scattering. Let the decay time for this process be ϕ. ϕ is a molecular property and is not a function of temperature and pressure. Process 2 leads to absorption, which has, as its inverse, thermal emission. Let the thermalization time be $\eta(p)$. As discussed in §1.5.1, an important aspect of the thermalization time is that it is inversely proportional to the pressure.

The *rates* of scattering and thermalization are ϕ^{-1} and η^{-1}, respectively. We may look upon this problem as if the single-scattering albedo, equation (3.5), were,

$$a_\nu = \frac{\phi^{-1}}{\phi^{-1} + \eta^{-1}(p)}.$$

From equation (4.8), the source function is,

$$J_\nu = B_\nu \frac{\eta^{-1}(p)}{\phi^{-1} + \eta^{-1}(p)} + J_\nu(scattering) \frac{\phi^{-1}}{\phi^{-1} + \eta^{-1}(p)} \ . \quad (4.21)$$

The source function (4.21) behaves very differently in the two limits $\eta \ll \phi$ and $\eta \gg \phi$. In the first limit, $J_\nu \to B_\nu$, and the equilibrium source function of Kirchoff and Planck is valid. This is a case of rapid collisions, appropriate to high pressures. In the second limit we have a scattering source function, with no transfer of energy to or from the thermal reservoir. This is the case of slow collisions, appropriate to low pressures. The transition between these two different behaviors occurs when $\phi = \eta(p)$.

An important practical question in atmospheric thermodynamics is the source function for the 15 μm band of carbon dioxide. This is the band that is responsible for most of the thermal cooling rate above 20 km (see Figure 2.4). The band arises from transitions between vibrational levels and has a natural lifetime $\phi = 0.74$ s. The thermalization time, η, has been measured in the laboratory to be 25 μs at 10^5 Pa and 180 K. The level at which $\phi = \eta(p)$ occurs at a pressure of 3.4 Pa, or an altitude near 76 km. Below this level the equilibrium source function should be used while, at higher levels, the band will scatter radiation and will have relatively little effect on the thermodynamics of the atmosphere.

For the sake of completeness we may discuss the form of $J_\nu(scattering)$ to use in equation (4.21). There are more possibilities for scattering source functions than equation (4.15). (4.15) is the result of a *frequency-coherent process*, a process for which the scattered photon has the same frequency as the incident photon, and we have seen that this scattering leads to monochromatic radiative equilibrium. This is not appropriate to the present circumstance. While the vibrational levels depart from thermal equilibrium at pressures below 3.4 Pa, rotational levels remain in equilibrium to much lower pressures, because they are more readily perturbed by collisions, see §1.4.2 and §1.4.3. Thus, an absorbed quantum may be emitted with the same vibrational energy, but with a different rotational energy, in the same vibration-rotation band, but at a different frequency. Integrated over the entire vibration band input and output must still be equal, but not for any single frequency. The following *incoherent scattering* source function fits this prescription,

$$J_\nu(scattering) = B_\nu \frac{\int_{band} e_\nu \, d\nu \int_{4\pi} I_\nu \, d\omega}{\int_{band} e_\nu \, d\nu \int_{4\pi} B_\nu \, d\omega}. \quad (4.22)$$

Equation (4.22) satisfies the radiative equilibrium equation, (4.7), not for each frequency independently, but only when a frequency integral is performed over the 15-μm band. This demonstrates that (4.22) is a plausible source function, but its precise form depends on details of the process.

RADIATIVE TRANSFER

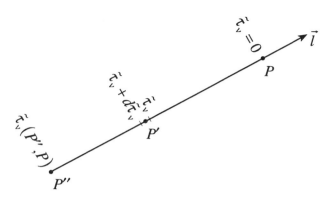

Figure 4.4 Path of integration.

4.2 Solutions to transfer problems

4.2.1 The integral equation

The equation of transfer may be transformed into an integral equation, with the source function appearing in the integrand. For thermal source functions the integrand is defined by the atmospheric state, which may be regarded as given for the radiation problem. This makes the integral the solution of choice for those numerical weather and climate models that use only thermal source functions.[4] For multiple scattering, however, the integrand contains the radiance that is to be calculated, and the resulting integral equation presents mathematical difficulties. The method that we shall describe in §4.2.2 is preferred for scattering source functions.

The *optical path* between two points is defined by, see §3.1,

$$\tilde{\tau}_\nu(1,2) = \int_1^2 n e_\nu \, dl \;. \tag{4.23}$$

$\tilde{\tau}_\nu$ is dimensionless and positive definite.

The geometry of the path of integration is shown in Figure 4.4. $\tilde{\tau}_\nu$ is defined to be zero at the point P, where the radiance is to be evaluated. The equation of transfer at P' for a thermal source function is, see (4.4),

$$\frac{dI_\nu(P', \vec{l})}{d\tilde{\tau}_\nu} = I_\nu(P', \vec{l}) - B_\nu(P') \;. \tag{4.24}$$

---

[4]This is not to suggest that any weather and climate models take no account at all of scattering. Clouds and ground can, with some justification, be treated as black surfaces in the thermal spectrum, and assigned a surface reflectivity or albedo in the solar spectrum.

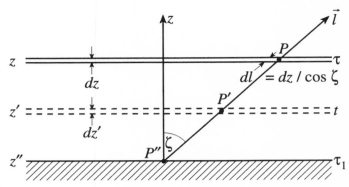

Figure 4.5 Geometry of a stratified atmosphere. ζ is the zenith angle. τ, t, and τ_1 are optical depths. For a downward-directed radiance, P' is higher than P, and $\tau > t$.

The sign of equation (4.24) differs from that of (4.4) because a displacement of P' in the l-direction decreases $\tilde{\tau}_\nu$.

The formal solution to equation (4.24) is,

$$I_\nu(P, \vec{l}) = I_\nu(P'', \vec{l}) \exp[-\tilde{\tau}_\nu(P'', P)] + \int_0^{\tilde{\tau}_\nu(P'', P)} B_\nu(P') \exp[-\tilde{\tau}_\nu(P', P)]\, d\tilde{\tau}_\nu. \tag{4.25}$$

Radiation arriving at P consists of radiation originating from a boundary at P'', $I_\nu(P'', \vec{l})$, attenuated by the optical path between P and P'', plus contributions from elements of the path in between, each attenuated by the appropriate optical path.

Equation (4.25) may be written in a more compact form by allowing the upper limit to the integral to be infinity,

$$I_\nu(P, \vec{l}) = \int_0^\infty B_\nu(P') \exp[-\tilde{\tau}_\nu(P, P')]\, d\tilde{\tau}_\nu . \tag{4.26}$$

Formally this requires that we replace the source function in the interval $\tilde{\tau}_\nu(P'', P) \leq \tau_\nu < \infty$ by the surface radiance, $I_\nu(P'', \vec{l})$.

The geometry for a stratified atmosphere is shown in Figure 4.5. The reference direction is vertical and the angular variable is $\xi = \xi_z(\vec{l}) = \cos\zeta$, where ζ is the zenith angle.

The appropriate independent variable for a stratified atmosphere is the *optical depth* (no tilde), which is the vertical optical path from the level of interest to space,

$$\tau_\nu(z) = \int_z^\infty e_\nu(z') n(z')\, dz' . \tag{4.27}$$

RADIATIVE TRANSFER

From Figure 4.5,
$$\tilde{\tau}_\nu(P, P') = \frac{[\tau_\nu(z') - \tau_\nu(z)]}{\xi} . \tag{4.28}$$

We now substitute (4.28) into (4.25) and apply the boundary conditions. These boundary conditions differ according to whether the radiance is directed upwards or downwards, and we distinguish between the two radiation streams with the notations I_ν^+ and I_ν^-. There is effectively no thermal radiation from space,
$$I_\nu^-(\tau = 0, \xi) = 0, \quad \text{for } -1 \leq \xi < 0 . \tag{4.29}$$

For upward-travelling radiation it is common practice to assume that the surface emits as a black body at temperature, T_g. T_g is the *emission temperature*, defined as the temperature that gives the required surface radiance from the Stefan-Boltzmann law, equation (4.13); it may not be exactly the same as the ground temperature. With this definition,
$$I_\nu^+(\tau_1, \xi) = B_\nu(T_g), \quad \text{for } +1 \geq \xi > 0 , \tag{4.30}$$
where τ_1 is the optical depth at the surface.

With these boundary conditions, and the definitions of t, and τ given in Figure 4.5,
$$\begin{aligned} I_\nu^+(\tau, \xi) &= B_\nu(T_g) \exp\left(-\frac{\tau_1 - \tau}{\xi}\right) \\ &+ \int_\tau^{\tau_1} B_\nu(t) \exp\left(-\frac{t - \tau}{\xi}\right) \frac{dt}{\xi}, \quad \text{for } +1 \geq \xi > 0 , \end{aligned} \tag{4.31}$$

and,
$$I_\nu^-(\tau, \xi) = -\int_0^\tau B_\nu(t) \exp\left(-\frac{t - \tau}{\xi}\right) \frac{dt}{\xi}, \quad \text{for } -1 \leq \xi < 0 . \tag{4.32}$$

Finally, we may evaluate the vertical flux of radiation from equations (4.5), (4.31), (4.32), and Figure 4.3,
$$\begin{aligned} F_\nu(\tau) &= -\int_{+1}^{-1} 2\pi I_\nu(\tau, \xi) \xi \, d\xi, \\ &= 2\pi B_\nu(T_g) E_3(\tau_1 - \tau) + 2\pi \int_{E_3(\tau_1 - \tau)}^{1/2} B_\nu(t) \, dE_3(t - \tau) , \\ &\quad -2\pi \int_{E_3(\tau)}^{1/2} B_\nu(t) \, dE_3(\tau - t) , \end{aligned} \tag{4.33}$$
where E_3 is a tabulated function called the third exponential integral,
$$E_3(x) = \int_0^1 y \exp\left(-\frac{x}{y}\right) dy . \tag{4.34}$$

Some comments may save the reader difficulties with signs when reading the literature. First, radiances, source functions, and optical depths are all positive definite quantities. Second, exponential attenuating factors must be less than unity. Third, up to this point we have preserved the vectorial character of ξ and F; they are both positive upwards and negative downwards. For a stratified atmosphere, however, upward and downward flux components have different boundary conditions and must be calculated independently. Consequently, a nonvectorial notation with only positive flux components is convenient. Equation (4.33) may be written,

$$F_\nu(\tau) = F_\nu^+(\tau) - F_\nu^-(\tau) \,, \tag{4.35}$$

where F^+ is equal to the first two terms on the right of equation (4.33), while F^- is equal to the magnitude of the third term. The angular variable may also be redefined,

$$\mu = |\xi| \,. \tag{4.36}$$

Since the upward and downward fluxes are identified with appropriate superscripts, as in (4.35), this redefinition is unambiguous. For example, solar radiation is always directed downwards, and $\xi_\odot$ (the cosine of the solar zenith angle) is always negative. Equations (4.19) and (4.27) give for the solar irradiance,

$$f_\nu(\tau) = f_{\nu,0} \exp\left(-\frac{\tau_\nu}{\mu_\odot}\right) \,. \tag{4.37}$$

4.2.2 Doubling and adding

In this section we discuss one of a number of techniques for calculating radiative quantities in a stratified atmosphere, without placing restrictions on the source function. Different techniques have been developed to take advantage of fast algorithms available on modern computers. All are capable of essentially unlimited accuracy, and the only issue is speed and cost. The *doubling-and-adding method* is perhaps the simplest to describe. We shall give only an abbreviated version of this approach, although it makes all the essential points.

We first establish the *interaction principle*, a statement of energy conservation and of Lambert's law that replaces the equation of transfer for discrete, stratified layers. Figure 4.6 shows the configuration of radiances at the boundaries of a discrete layer of optical thickness τ_a, lying between optical depths τ_1 and τ_2. Radiances impinging on this layer are $\mathbf{I}^-(\tau_1)$ from above and $\mathbf{I}^+(\tau_2)$ from below; the outward directed radiances, $\mathbf{I}^+(\tau_1)$ and $\mathbf{I}^-(\tau_2)$, are each sums of three outward components shown in Figure 4.6.

The radiances have been written in bold face because each represents an array of radiances, each radiance with different, discrete values of the

RADIATIVE TRANSFER

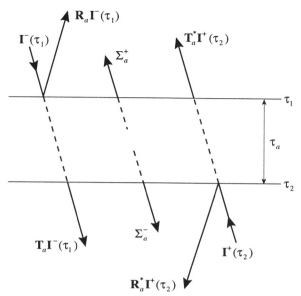

Figure 4.6 The interaction principle. The figure illustrates definitions of quantities introduced in the text.

zenith and azimuth angles.[5] The angles are chosen in a manner convenient for evaluating the flux integral, equation (4.5). For the zenith angle, Gauss' method is often used, whereby values of μ are chosen to equal zeros of a Lagrange polynomial. Azimuth angle, on the other hand, is often handled by Fourier decomposition. The nature of the choice does not affect our discussion, and any level of accuracy may be obtained by using a sufficiently large number of ordinates.

We now make an assertion that is physically equivalent to Lambert's law: that the outward-directed radiances are linearly related to the inward-directed, or incident radiances; in addition, there will be contributions to the outward radiances from internal sources, represented in Figure 4.6 by the quantities $\Sigma_a^{+,-}$. The internal sources can be thermal or primary scattering from the solar beam, or both. Thus the incident solar beam is an external parameter to the problem. The quantities calculated by this method correspond to the scattered radiance, I_s, that was discussed in §4.1.4. To avoid overburdening symbols with suffixes, we shall omit the s-suffix as well as ν-suffixes from the following equations. The reader must bear in mind that, in the end, only diffuse radiances and flux components will have been calculated by the doubling-and-adding method and that the

---

[5] In contrast to the previous section, we need to specify both μ and the azimuth angle, ϕ, in scattering problems.

direct solar beam must be added for the complete solution.

The coefficients in these linear relationships are the *reflection* and *transmission functions*, $\mathbf{R}_a, \mathbf{R}_a^*, \mathbf{T}_a$, and $\mathbf{T}_a^*$. For definitions of these quantities look at Figure 4.6. The following two equations express the interaction principle: for energy conservation the outgoing radiances are each formed from the sum of three terms,

$$\begin{aligned} \mathbf{I}^+(\tau_1) &= \mathbf{R}_a \mathbf{I}^-(\tau_1) + \mathbf{T}_a^* \mathbf{I}^+(\tau_2) + \mathbf{\Sigma}_a^+ , \\ \mathbf{I}^-(\tau_2) &= \mathbf{T}_a \mathbf{I}^-(\tau_1) + \mathbf{R}_a^* \mathbf{I}^+(\tau_2) + \mathbf{\Sigma}_a^- , \end{aligned} \quad (4.38)$$

or, in matrix notation,

$$\begin{pmatrix} \mathbf{I}^+(\tau_1) \\ \mathbf{I}^-(\tau_2) \end{pmatrix} = \mathbf{S}(a) \begin{pmatrix} \mathbf{I}^-(\tau_1) \\ \mathbf{I}^+(\tau_2) \end{pmatrix} + \begin{pmatrix} \mathbf{\Sigma}_a^+ \\ \mathbf{\Sigma}_a^- \end{pmatrix} , \quad (4.39)$$

where

$$\mathbf{S}(a) = \begin{pmatrix} \mathbf{R}_a & \mathbf{T}_a^* \\ \mathbf{T}_a & \mathbf{R}_a^* \end{pmatrix} . \quad (4.40)$$

Now consider the combination of two layers to form a single layer. The process is illustrated in Figure 4.7. The algebra is straightforward but tedious, and we shall present only a summary. The second layer has an optical thickness, τ_b, and lies between optical depths τ_2 and τ_3; it obeys the relation (4.38) with a-suffixes replaced by bs. Since the layer a lies above b, these two equations have the radiances $\mathbf{I}^+(\tau_2)$ and $\mathbf{I}^-(\tau_2)$ in common, and these two quantities may be eliminated to yield relationships between radiances at τ_1 and τ_3. The formal equivalent of equation (4.39) is,

$$\begin{pmatrix} \mathbf{I}^+(\tau_1) \\ \mathbf{I}^-(\tau_3) \end{pmatrix} = \mathbf{S}(a+b) \begin{pmatrix} \mathbf{I}^-(\tau_1) \\ \mathbf{I}^+(\tau_3) \end{pmatrix} + \begin{pmatrix} \mathbf{\Sigma}_{a+b}^+ \\ \mathbf{\Sigma}_{a+b}^- \end{pmatrix} . \quad (4.41)$$

It is a straightforward exercise in matrix algebra to express the unknown quantities, $\mathbf{S}(a+b)$ and $\mathbf{\Sigma}_{a+b}^{+,-}$, in terms of the reflection and transmission coefficients for the individual layers a and b, but we shall not do so here. The reader is referred to books and documents given in §4.4, Reading, for details. It is sufficient to note that the expressions can be derived.

We may now look at the two layers as a single combined layer, c, to which equation (4.37) must apply with a-suffixes replaced by c-suffixes; $\mathbf{S}(a+b) = \mathbf{S}(c)$, and $\mathbf{\Sigma}_{a+b}^{+,-} = \mathbf{\Sigma}_c^{+,-}$. Omitting once again the matrix algebra, relationships between the reflection and transmission functions for the combined layer and those for the component layers are,

$$\begin{aligned} \mathbf{R}_c &= \mathbf{R}_a + \mathbf{T}_a^*(\mathbf{I} - \mathbf{R}_b \mathbf{R}_a^*)^{-1} \mathbf{R}_b \mathbf{T}_a , \\ \mathbf{T}_c &= \mathbf{T}_b(\mathbf{I} - \mathbf{R}_a^* \mathbf{R}_b)^{-1} \mathbf{T}_a , \\ \mathbf{R}_c^* &= \mathbf{R}_b + \mathbf{T}_b(\mathbf{I} - \mathbf{R}_a^* \mathbf{R}_b)^{-1} \mathbf{R}_a^* \mathbf{T}_b^* , \\ \mathbf{T}_c^* &= \mathbf{T}_a^*(\mathbf{I} - \mathbf{R}_b \mathbf{R}_a^*)^{-1} \mathbf{T}_b^* . \end{aligned} \quad (4.42)$$

RADIATIVE TRANSFER

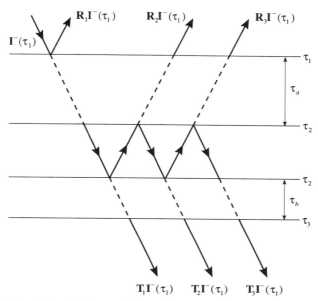

Figure 4.7 Combining two layers. The two layers a and b may be combined and treated as a single, combined layer, c.

I (with no indices) is the identity matrix. There is an analogous expression for $\mathbf{\Sigma}_c^{+,-}$.

These equations provide the basis for a complete solution to any problem in a stratified atmosphere. Let us suppose that we know the transmission and reflection functions for a number of layers, each of finite optical thickness, that make up a stratified atmosphere. The equation enables us to combine them two at a time until we have constructed the reflection and transmission functions for the entire atmosphere. Once this solution has been obtained, boundary conditions may be applied, and then the solution may be taken apart layer by layer to yield data at all interfaces between layers. This is the *adding method*. It remains to discuss the optical properties of the individual layers.

Numerical models of the atmosphere are always divided up into finite, homogeneous layers—the fewer the better. The number of layers may be set by the limited observational data or by other requirements external to the radiation problem. In any event, the layers are unlikely to be optically thin ($\tau_a \ll 1$), which is the only circumstance under which the properties of a layer can be calculated without solving a transfer equation. We may, however, construct the properties of an optically thick layer from those of an optically thin layer by using the *doubling method*. We may illustrate the doubling method by calculating the coefficients in the matrix $\mathbf{R}_a$. This matrix relates upward scattered radiation to downward incident radiation at the top of a scattering layer, see the upper left of Figure 4.6.

Let the layer be optically thin, $\tau_a = \Delta\tau \ll 1$. For an optically thin layer the emitted radiance is given by equation (4.3),

$$I^+(\mu', \phi') = J(\mu', \phi') sn\, dl \,. \qquad (4.43)$$

We have replaced the vector $\vec{l}$ by (μ', ϕ') where ϕ' is the azimuth angle.

The optical thickness of the layer is $\Delta\tau = en\, dz$, from equation (4.27). From Figure 4.5, and the definition of the single scattering albedo, equation (3.5), we have $sn\, dl = \frac{a\Delta\tau}{\mu'}$. In terms of μ and ϕ, the solid angle may be written, $d\omega = d\phi d\mu$, see Figure 4.3 for the geometry involved. If we bring all of these results together with the scattering source function, equation (4.15), the radiance scattered from the infinitesimal layer is,

$$I^+(\mu', \phi') = \frac{a\Delta\tau}{\mu'} \int_0^{2\pi} \int_0^1 I^-(-\mu, \phi) P(\mu', \phi'; -\mu, \phi) \frac{d\phi d\mu}{4\pi} \,. \qquad (4.44)$$

For numerical methods integrals are replaced by discrete sums,

$$\int_0^{2\pi} \int_0^1 X(\phi, \mu)\, d\phi\, d\mu \to \sum_{ij} X_{ij} c_{ij} \,. \qquad (4.45)$$

The choice of discrete coordinates (ij) and the coefficients c_{ij} depends upon the method of numerical integration that is to be used. As mentioned before, it is common practice to use gaussian subdivision for μ and Fourier decomposition for ϕ. Whatever the methods used, (4.44) and (4.45) may be combined into a relation,

$$\begin{aligned} I^+_{i'j'}(\Delta\tau) &= \frac{a\Delta\tau}{4\pi\mu_{i'}} \sum_{ij} I^-_{ij} P_{ij,i'j'} c_{ij}, \\ &= R_{ij,i'j'}(\Delta\tau) I^-_{ij} \,. \end{aligned} \qquad (4.46)$$

The sum rule for repeat indices is used in the last expression. The right side of (4.46) is equivalent to $\mathbf{R}_a \mathbf{I}^-$, and $R_{ij,i'j'}(\Delta\tau)$ are the elements of the matrix $\mathbf{R}_a$ for an infinitesimally thin scattering layer. There are analogous expressions for $\mathbf{T}_a$, $\mathbf{R}_a^*$, and $\mathbf{T}_a^*$.

We now proceed as follows. Apply equation (4.46) to a homogeneous layer of thickness $\Delta\tau = \tau_a \times 2^{-20}$. For all practical purposes this layer will be optically thin. Add two layers, using equation (4.42), to give the properties of a layer of optical thickness $2\Delta\tau$, and repeat the process for a layer of thickness $4\Delta\tau$. Twenty such operations yield the properties of the required atmospheric layer. This is the second component of the doubling-and-adding method. While the whole process may appear to be complicated, only linear operations, matrix multiplications, and inversions are involved, and these may be performed rapidly using standard algorithms that are commonly available on modern computers.

4.3 Approximate methods

Thoughtful approximations are the key to almost all atmospheric calculations, and this is particularly true of radiation calculations. While the methods described in §4.2 can lead, in principle, to complete numerical solutions, they are rarely used without approximation, and the reader should have some familiarity with the ideas involved.

4.3.1 The frequency integration

Up to this point we have presented equations for *monochromatic quantities*, with ν-suffixes, without dealing with the fact that the important thermodynamic results, for example, the flux divergence equations (4.6), involve an integration over all frequencies. How many monochromatic calculations must be made in order to evaluate this integral? A glance at the 4 cm^{-1} interval shown in Figure 3.3(b) suggests that this number may be very large. The number depends upon the number of lines, and on the line width, but with modern digital archives containing close to 10^6 lines, 10^7 to 10^8 frequency intervals is a plausible order-of-magnitude estimate.

It is possible to perform 10^7 to 10^8 calculations of the kind described in §4.2 with large computers; these are called *line-by-line calculations*, and may be used for establishing standard results. However, there is always internal competition for computing time in modern meteorological and climate calculations, and line-by-line calculations are not regarded as an appropriate use of limited resources. Economical ways to evaluate frequency integrals over line spectra are essential.

Radiative equations, such as the flux equation (4.33), contain two very different scales of frequency variation. We have mentioned the line-absorption spectrum, but there are also the scales of variation of the Planck function, the scattering coefficient, and the phase function. In practice, none of these three need be considered in great detail; a relatively small number of frequency intervals (~ 20) would suffice to integrate over all frequencies if these were the only frequency-dependent factors involved.

We may define frequency intervals, $\Delta\nu_i$, over which the Planck function, the phase function, and the scattering coefficient are approximately constant, and may without serious error be assigned mean values B_i, P_i, and s_i, respectively. We now consider how to evaluate directly the mean values over such intervals of quantities that depend upon the line absorption coefficients. Let Q_ν be any spectral quantity (radiance, flux, flux divergence, transmission, etc.); we wish to derive the average value,

$$Q_i = \int_0^1 Q_\nu \, d\left(\frac{\nu}{\Delta \nu_i}\right) . \tag{4.47}$$

The integrated quantity, Q, may be obtained from the sum,

$$Q = \int_0^\infty Q_\nu \, d\nu = \sum_i Q_i \Delta \nu_i . \tag{4.48}$$

Over the past 50 to 60 years much effort has gone into the effort to understand the behavior of the spectrally averaged functions, (4.47), particularly when Q_ν is the transmission $\mathcal{T}_\nu$, when they are referred to as *transmission functions*. These studies were partly theoretical and partly empirical, and reached a high level of sophistication; at this time (1995) they still form the basis of most numerical approaches to radiative transfer in the earth's atmosphere. They are, however, being superseded by methods that can make direct use of the numerical archives that were discussed in §3.2.1, and we shall describe only these more recent developments.

Consider the numerical evaluation of (4.47) by a line-by-line approach. The spectral interval must be divided into narrow subintervals in each of which k_ν is effectively constant. A radiation calculation is performed for each subinterval, using the appropriate k_νs for each layer, and a sum taken over the all the subintervals of the spectral interval. Assume, for the moment, that we are dealing with a homogeneous absorption path along which k_ν does not change. Each subinterval will then have a unique value of k, which determines the radiative quantity, Q_ν. It does not matter where in the interval this calculation is made, only the value of k is important; we may therefore refer to the calculated quantity as Q_k. If the fraction of the spectral interval having k_ν between k and $k + dk$ is $f(k) \, dk$, we may write (4.47) in an alternate form,

$$Q_i = \int_0^\infty Q_k f(k) \, dk . \tag{4.49}$$

From the numerical point of view, (4.49) and (4.48) are identical if corresponding ν- and k-subintervals are employed. The two calculations differ only in the ordering of the subintervals within the spectral interval, and this is unimportant. There is an important difference, however: while Q_ν is a rapidly varying function of ν, Q_k is a slowly varying function of k. Typically, 20 k-intervals give all the accuracy that is required for a numerical quadrature of equation (4.49). This reduction of effort by four to five orders of magnitude brings the calculation into the realm of feasibility for climate models. This is the *k-distribution method*. It has a long history in the astrophysics community. Note that we have not talked about absorption alone. The k-distribution method describes how to deal with that part of the problem that arises from the line structure. The same $f(k)$ applies to any radiation calculation.

Up to this point, a radiation calculation can be made as accurate as is desired, by choosing the number of k-intervals, but when we consider a nonhomogeneous atmosphere, approximation is unavoidable. $f(k)$ is a function of the temperature and pressure; it changes from atmospheric level to level, and is not unique, as assumed above. The essence of the problem is as follows. Frequency (or wavelength) is the natural physical variable for radiation problems. Radiation passing from one level to another does so at constant frequency, and the physics of absorption tells us how to calculate the absorption coefficient at each frequency. To solve a radiation problem, we select a frequency, calculate the absorption coefficient for each layer, and use this in a doubling-and-adding or similar procedure. This is the line-by-line method. In the k-distribution approach the spectrum is scrambled at each level, and this physical connection between levels is lost. How, then, is it possible to accommodate a k-distribution calculation with levels having different temperatures and pressures?

The technique that is now adopted is to employ a new variable in place of the frequency, the *cumulative k-distribution*,

$$g(k) = \int_0^k f(k')\, dk' . \tag{4.50}$$

From (4.49),

$$Q_i = \int_0^1 Q_g\, dg . \tag{4.51}$$

Comparison between (4.51) and (4.47) suggests an analogy between $\left(\frac{\nu}{\Delta \nu_i}\right)$ and g. Let us assume, for the moment, that this analogy exists, that a calculation at constant g gives the same results as a calculation at constant ν, and ask how the numerical calculation would proceed. At this point the proposition may strain credulity, but we shall return to the evidence for its validity.

First, note that, from equation (4.50), g is a monotonic function of k. For given g, there is, at each atmospheric level in the calculation, a specific value of $k = k_g$ corresponding to it. In general, k_g is different for each layer. The numerical procedure is now the same as for a line-by-line calculation but with g in place of ν. We select a number of gs at intervals suitable for integrating (4.51) by numerical quadrature. Again, this number is very small compared to the number of monochromatic frequency subintervals. Experience suggests that 10 to 20 g-intervals are sufficient for most purposes. For each g there is a value of k_g for each level. These values of the absorption coefficient may be used in the integrals (4.31) and (4.32), or in the doubling-and-adding method, or in any other numerical method, to calculate radiances, and hence any other radiative quantity. With these data, the integral (4.51) may be performed numerically. This is the *correlated-k method*.

We must now discuss the accuracy of the correlated-k method and the reason for this accuracy. The correlated-k method is a product of numerical analysis, and we reverse the normal order of things by discussing the accuracy before discussing the rationale. Numerical tests have been made that suggest that errors for one spectral interval are rarely greater than 1% for any radiative quantity, radiance, flux, flux divergence, etc. for either thermal or solar radiation. Since several spectral intervals must be summed to form a complete frequency integral, errors will normally be less than this figure. Calculations for a seven-layer, stratified atmosphere, with both absorption and scattering, are shown in Figure 4.8. These calculations are typical of many that have been made by different investigators.

Finally we come to the question of why, apart from the analogy between equations (4.47) and (4.51), should the correlated-k technique be satisfactory? Theoretical studies show that the procedure is exact if the absorption coefficients in each layer are perfectly correlated. They will be partially correlated, but never completely for real bands. More importantly, the correctness of the correlated-k technique can be demonstrated in a number of limiting cases, which between them cover so many possibilities that little room is left for errors.

One such situation is illustrated in Figure 4.9. This figure shows a single, isolated absorption line, represented here by a half-line; it is self-evident that, for this simple configuration,

$$dg \equiv \frac{d\nu}{\Delta \nu_i} \ . \qquad (4.52)$$

Another circumstance when the result (4.52) is exact is for a band consisting of equally spaced lines of equal line strength, the *Elsasser band model*. That this is so may be seen by recognizing that the Elsasser band is no more than Figure 4.9, first reversed, and then repeated along the frequency axis.

Two further circumstances involve the asymptotic behavior of bands with *any* spectral structure. If all the lines in an atmospheric layer are *weak*, that is, the line centers are all weakly absorbing, the procedure is exact. If all the lines in an atmospheric layer are *strong*, that is, the line centers are all opaque, the procedure is also valid. The weak line condition, and the single-line condition illustrated in Figure 4.9, apply to *any* line shape. These circumstances greatly restrict the errors that can occur with real bands but, in the final analysis it is the numerical evidence of Figure

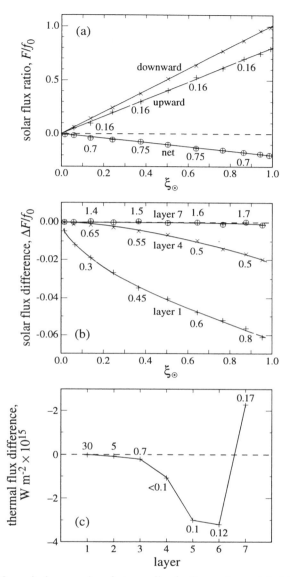

Figure 4.8 Numerical comparison between line-by-line and correlated-k computations. Calculations are presented for the frequency interval 5000 to 5050 cm^{-1} in a CO_2 band. The atmospheric model consists of seven homogeneous layers with individual pressures from 0.9 to 0.05 bar and temperatures from 200 to 295 K. Three of the layers contain scattering material in addition to CO_2. f_0 is the solar irradiance above the atmosphere. The top two panels are solar data and the bottom panel is thermal. The top panel refers to the flux at the top of the atmosphere. The term *flux difference* refers to the difference of flux between the top and the bottom of a layer. There are scarcely any differences between the points ($\times, +, \oplus$) and the lines. The points are the correlated-k data, while the line segments join the line-by-line data. The differences between the two are given as percentages by the numbers shown against some of the points.

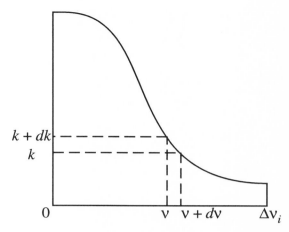

Figure 4.9 $\nu \to g$ mapping for a single line. $f(k)\,dk = dg$ is the fraction of the ν-axis corresponding to absorption coefficients between k and $k+dk$. The figure demonstrates that, for an isolated line, dg and $\frac{d\nu}{\Delta\nu_i}$ are identical.

4.8 that is important.

4.3.2 Approximate differential equations

Many of the formal difficulties involved in solving the transfer equation are associated with the detail of the angular distribution of the radiation field. If we are interested in angle-averaged quantities, such as the radiation flux, it is reasonable to suppose that we might take some liberties with the angular properties of the radiation field without changing the essential physics of the problem. Even the lowest order approximations to the angular distribution of the radiation field lead to results that can be remarkably accurate.

The *method of moments* employs the moment operator $\int_{4\pi}[\ \ ]\xi^m\,d\omega$ where [ ] is the argument of the operator, ω is the solid angle, and m is the order of the moment. If we change notation for the angular variables from $\vec{l} \to (\xi, \phi)$ and $\vec{d} \to (\xi', \phi')$, and omit ν-suffixes, we may write the equation of transfer for a stratified atmosphere in terms of the optical depth from (4.4), (4.27), and (4.28),

$$\xi \frac{dI(\xi,\phi)}{d\tau} = I(\xi,\phi) - J(\xi,\phi) , \qquad (4.53)$$

where, from (4.20),

$$J(\xi,\phi) = (1-a)B + a\int_{4\pi} I(\xi',\phi')P(\xi,\phi;\xi',\phi')\frac{d\omega'}{4\pi} + af\frac{P(\xi,\phi;\xi_\odot,\phi_\odot)}{4\pi} . \qquad (4.54)$$

RADIATIVE TRANSFER

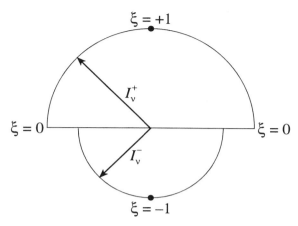

Figure 4.10 A semi-isotropic field of radiation.

Following the discussion of §4.1.4, the direct solar radiance is not included in $I(\xi', \phi')$, which corresponds to $I_{\nu,d}$ in that discussion.

When the moment operator is applied to the equation of transfer, we obtain a relationship between two moments whose orders differ by unity. Starting from $m = 0$, a series of equations can be built up with increasing order. These equations may be closed by assuming that a moment of high order can be approximated in terms of a moment of lower order.

If we apply the moment operator to the radiance itself, the zero and first moments ($m = 0, 1$) are,

$$\int_{4\pi} I \, d\omega = 4\pi \overline{I}, \tag{4.55}$$

and

$$\int_{4\pi} \xi I \, d\omega = F, \tag{4.56}$$

where $\overline{I}$ is the *mean radiance*, and F is the flux.

The second moment of the radiance may be calculated on the assumption of a *semi-isotropic field of radiation*, for which I^+ and I^- are individually constant but not equal to each other, see Figure 4.10,

$$\int_{4\pi} \xi^2 I \, d\omega \approx \frac{4\pi}{3} \overline{I}. \tag{4.57}$$

This approximate relationship between moments can be used for closure.

The zero moment of (4.53) is, from (4.5),

$$\frac{dF}{d\tau} = 4\pi(1-a)(\overline{I} - B) - af, \tag{4.58}$$

and, from (4.57), the first moment is,

$$\frac{4\pi}{3}\frac{d\overline{I}}{d\tau} = F(1-ag) - ag\xi_\odot f \ . \tag{4.59}$$

The derivation of (4.59) requires one theorem and one definition. The theorem, which we shall not prove, is,

$$\int_{-1}^{+1} P(\xi,\phi;\xi',\phi')\xi\,d\xi = g\xi' \ , \tag{4.60}$$

where (the definition),

$$g = \int_{4\pi} P(\cos\theta)\cos\theta\,\frac{d\omega}{4\pi} \ . \tag{4.61}$$

θ is the angle of scattering, and g is the *asymmetry factor* that defines the general nature of the scattering phase function,

$$g = \left\{ \begin{array}{rl} +1 & forward\ scattering, \\ 0 & isotropic\ scattering, \\ -1 & backward\ scattering. \end{array} \right\} \ . \tag{4.62}$$

To simplify the algebra, we now assume that a and g are both constants. Eliminate $\overline{I}$ between (4.58) and (4.59) and use equation (4.19),

$$\frac{d^2 F}{d\tau^2} = 3(1-a)(1-ag)F - 4\pi(1-a)\frac{dB}{d\tau} - f\left[\frac{a}{\xi_\odot} + 3ag\xi_\odot(1-a)\right] \ . \tag{4.63}$$

Equation (4.63) is for the diffuse flux only. The direct solar flux must be added,

$$F_\odot = \xi_\odot f \ . \tag{4.64}$$

Equation (4.63) is an ordinary, second-order differential equation with constant coefficients that may be solved by standard techniques if the boundary fluxes are known. Solutions may be obtained in terms of known functions, providing a valuable tool for preliminary investigations. The approximations that have been made are numerical rather than physical, and the solutions correctly reflect the physical properties of the problem.

In §4.2.1 the boundary conditions were placed on the radiance rather than the flux. At the upper boundary we know I^-, while at the lower boundary we know I^+. In order to turn these into boundary conditions on the flux, we assume a semi-isotropic radiation field. From Figure 4.10,

$$\overline{I} = \frac{I^+ + I^-}{2} \ , \tag{4.65}$$

RADIATIVE TRANSFER   115

and from the definition of flux, equation (4.5),

$$F = \pi I^+ - \pi I^- . \tag{4.66}$$

Either I^+ or I^- may be eliminated between (4.65) and (4.66). For example, at the lower boundary, $I^+ = B_g$ and,

$$F = 2\pi(B_g - \overline{I}) . \tag{4.67}$$

Equation (4.58) may then be used to eliminate $\overline{I}$ and to give the appropriate flux boundary condition. Details of this and similar boundary conditions are left to the reader.

4.3.3 Radiation to space

This approximation is valid only for a thermal source function and for $e = k$. From equations (2.1), (4.6), and the geometry of Figure 4.3, the heating rate per unit volume, expressed for a single frequency, is,

$$\rho \dot{q}_\nu = \int_{-1}^{+1} 2\pi n k_\nu (I_\nu - B_\nu) \, d\xi . \tag{4.68}$$

Equation (4.68) can be evaluated when the radiance is known. Expressions for the upward and downward radiances are given by equations (4.31) and (4.32). The *radiation-to-space approximation* exploits simplifications to these expressions that are possible at levels for which the optical depth is not large.

If the atmosphere above a specific level is translucent, that level can readily emit radiation to space, from which it receives no radiation in return. This gives rise to a very large cooling rate. Other contributions to the heating rate, both positive and negative, will arise from exchange of radiation with other atmospheric levels or with the ground. Since atmospheric temperatures in the lower and middle atmospheres are usually within ±20 K of the mean, the exchange terms with other levels involve a degree of cancellation. The radiation-to-space approximation neglects these exchange terms.

We may extract the term involving radiation to space by the artifice of assuming that all levels of the atmosphere, and the ground, have the same temperature as the level of interest, so that all exchange, except with space, is identically zero. With this assumption, (4.31) and (4.32) become,

$$I_\nu^+(\tau_\nu, \xi) = B_\nu(\tau_\nu), \tag{4.69}$$

$$I_\nu^-(\tau_\nu, \xi) = B_\nu(\tau_\nu)\left[1 - \exp\left(\frac{\tau_\nu}{\xi}\right)\right], \tag{4.70}$$

and, substituting in equation (4.68),

$$\begin{aligned}\rho \dot{q}_\nu &= -\int_{-1}^{0} 2\pi n k_\nu B_\nu(\tau_\nu) \exp\left(\frac{\tau_\nu}{\xi}\right) d\xi \\ &= -2\pi n k_\nu B_\nu(\tau_\nu) E_2(\tau_\nu), \end{aligned} \quad (4.71)$$

where

$$E_2(x) = \int_0^1 \exp\left(-\frac{x}{y}\right) dy, \quad (4.72)$$

is the second exponential integral, another tabulated function.

Equation (4.71) is supported by numerical evidence. Early calculations for the troposphere are shown in Figure 4.11, in which the radiation-to-space term is compared to an exact calculation using the same data for both calculations. In most cases there is surprisingly little difference. The reason why is subtle.

Compare heating rates for different frequencies with different optical depths, treating the optical depth as an independent variable. Equation (4.71) is analogous to the Chapman function that will be discussed in §5.1.4. If the absorption coefficient is constant, and if the number density of absorbers follows a barometric law, the Chapman function is shown to have a maximum for $\tau_\nu = 1$. Now, the line-absorbing character of atmospheric absorbers assures that, at every level, optical depths run the gamut from small to large, and there are always spectral regions in which $\tau_\nu \approx 1$. Provided that these regions are not too few, they dominate the heating, and (4.71) is a good approximation.

A further approximation is possible if equation (4.71) is used for perturbation calculations, as is commonly done in dynamical meteorology. If the temperature is perturbed at fixed τ_ν,

$$\delta(\rho \dot{q}_\nu) = -2\pi n E_2(\tau_\nu) \left(\frac{\partial k_\nu B_\nu(\tau_\nu)}{\partial T}\right)_0 \delta T. \quad (4.73)$$

The partial differential is evaluated at a mean temperature indicated by the subscript 0. Equation (4.73) is a statement of Newton's law of cooling, and is called the *Newtonian approximation*.

4.4 Reading

For discussion of radiative transfer in general, see Houghton (1986), Chapter 2, Liou (1980), and Goody and Yung (1990), Chapter 3. Each book contains many references. Houghton's book is the most elementary.

The doubling-and-adding method is treated in the following review paper

Hansen, J.E., and Travis, L.D., 1974, "Light scattering in planetary atmospheres," *Space Science Reviews* **16**, p. 527.

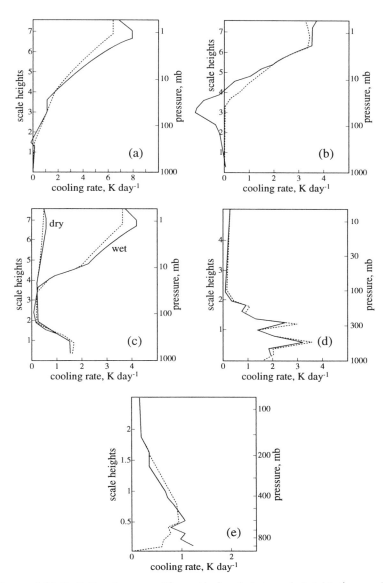

Figure 4.11 Radiation to space. The vertical scale is in scale heights (one scale height is between 6 and 8.5 km). (a) is for the 15 μm CO_2 band in mid-latitudes. (b) is for the 9.6 μm band of O_3 in the tropics. (c), (d), and (e) are for water vapor in mid-latitudes, the tropics, and polar regions, respectively. The full lines are exact calculations, while the dotted lines are for the radiation-to-space approximation.

4.5 Problems

Asterisks* and double asterisks** indicate higher degrees of difficulty.

4.1 Thermal emission from an isothermal, nonblack cloud, I.
In this and the following five questions we explore the optical properties of clouds using the simplest approximation to the radiative transfer equation, equation (4.63). This equation applies to a stratified atmosphere, so the cloud must also be stratified, a reasonable approximation for a stratus cloud. The top of the cloud is at $\tau = 0$, and the bottom is at $\tau = \infty$. For this problem calculations are required at the top of the cloud only.

First, assume that we are in the thermal region of the spectrum ($f = 0$), that the cloud is isothermal ($\frac{dB}{d\tau} = 0$), that the scattering is isotropic, ($g = 0$), and that the single-scattering albedo, a, is a constant. With these simplifications equation (4.63) becomes,

$$\frac{d^2 F}{d\tau^2} = \alpha^2 F, \ \alpha^2 = 3(1-a).$$

(i) Show that the boundary condition at $\tau = 0$ is,

$$\left(\frac{dF}{d\tau}\right)_{\tau=0} = 4\pi(1-a)\left(\frac{F(0)}{2\pi} - B\right).$$

Show that,
$$\frac{F(0)}{\pi B} = \frac{4(1-a)}{2(1-a) + \sqrt{3(1-a)}}.$$

(ii) The above formulae are approximate. What should the flux be for a black body ($a = 0$)? Correct the above formula with a constant factor to allow for this error and calculate $\frac{F(0)}{\pi B}$ for $a = 1.0, 0.8, 0.6, 0.4, 0.2, 0.0$.

(iii) It is normal practice to assume that a cloud emits as a black body whose surface has the same temperature as the cloud. Discuss this proposition in the light of the data in Figure 8.10.

4.2 Thermal emission from an isothermal, nonblack cloud, II.
Extend the treatment of Problem 4.1 by calculating the upward flux for the same values of a, but with $g = 0.5, 1.0$. Comment upon the singular case $g = 1$.

4.3 Thermal emission from an isothermal, nonblack cloud, III.* Extend the treatment of Problem 4.2 to a finite cloud of optical

RADIATIVE TRANSFER

depth, τ_1. Beneath the cloud is a black surface with a surface emission B_0 that is different from the Planck function of the cloud, B. Show that:

$$F(\tau) = Le^{+\alpha\tau} + Me^{-\alpha\tau}$$

$$L = \frac{-4\pi(1-a)\{Be^{-\alpha\tau_1}c_1 + (B_0 - B)c_2\}}{c_1^2 e^{-\alpha\tau_1} - c_2^2 e^{+\alpha\tau_1}},$$

$$M = \frac{-4\pi(1-a)\{Be^{+\alpha\tau_1}c_2 + (B_0 - B)c_1\}}{c_1^2 e^{-\alpha\tau_1} - c_2^2 e^{+\alpha\tau_1}},$$

where

$$c_1 = \alpha - 2(1-a),$$
$$c_2 = \alpha + 2(1-a),$$
$$\alpha^2 = 3(1-a)(1-ag).$$

4.4 Solar reflectivity or cloud albedo.* In this problem we consider the solar region of the spectrum for which $B = 0$ and, from (4.35),

$$f(\tau) = f(0)e^{-\tau/\mu_\odot}.$$

From (4.60), the direct solar flux is,

$$F_\odot = -\mu_\odot f.$$

(i) What is the approximate transfer equation for isotropic scattering, $g = 0$? And what is the upper boundary condition?

(ii) The cloud is infinitely thick and the upper boundary is at $\tau = 0$. Obtain an expression for the flux of scattered radiation as a function of optical depth.

(iii) Calculate the cloud albedo,

$$A = -\frac{F(\tau=0)}{F_\odot(\tau=0)},$$

for $\mu_\odot = 2/3$ and for $a = 1.0000, 0.9990, 0.9900, 0.9000, 0.7000, 0.0000$.

4.5 Line formation in a cloudy atmosphere. An important technique for obtaining information about remote planetary atmospheres is to analyze absorption lines in sunlight scattered back from their clouds. (Venus and the outer planets are completely cloud covered.) The problem involves a mixture of cloud particles with an absorbing gas. The optical properties of the cloud itself vary slowly with frequency, while a gaseous

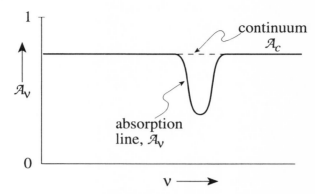

Figure 4.12 A planetary reflection spectrum.

vibration-rotation line varies rapidly over its narrow width. Thus, in a reflected spectrum, the absorption line appears as a sharp feature superimposed on a cloud continuum, see Figure 4.12. A measurable quantity is the "absorption," $\frac{A_c - A_\nu}{A_c}$, where A is the albedo.

(i) Show that, for an infinitely thick cloud with $g = 0$, the "absorption" is given by,

$$\frac{A_c - A_\nu}{A_c} = 1 - \frac{a_\nu}{a_c} \frac{\left\{1 + \frac{2}{3}\sqrt{3(1-a_c)}\right\}\left\{1 + \mu_\odot\sqrt{3(1-a_c)}\right\}}{\left\{1 + \frac{2}{3}\sqrt{3(1-a_\nu)}\right\}\left\{1 + \mu_\odot\sqrt{3(1-a_\nu)}\right\}}.$$

(ii) If the cloud scatters but does not absorb, and the molecules absorb but do not scatter, show that,

$$1 - a_\nu = \frac{k_{\nu,m} n_m}{s_p n_p + k_{\nu,m} n_m},$$

where the suffix m refers to molecules and the suffix p to particles. Hence obtain an expression for the "absorption" in the limit $(1 - a_\nu) \to 0$. Comment upon the qualitative difference between this result and that for a thin atmosphere without clouds, when scattering takes place from the planet's surface.

4.6 Direct and diffuse solar radiation. In Problem 4.4, three components of the solar flux were identified: the flux due to the direct, "parallel" solar beam, $F_\odot$, and two scattered or diffuse fluxes. If $(1-a) \ll 1$, as is usually the case in clouds, one of these fluxes has a much longer range than the other (which is which?). We may therefore speak of a short-range, diffuse flux, F_s, and a long-range, diffuse flux, F_l.

RADIATIVE TRANSFER

(i) In the limit $(1-a) \ll 1$, obtain expressions for the three flux components.

(ii) In the visible spectrum, $(1-a)$ for a water cloud is 10^{-3}. With $\mu_\odot = \frac{2}{3}$, calculate the three fluxes at $\tau = 0$. Which component is closest to the net flux?

(iii) With the same data as for (ii), at what optical depth are the two diffuse fluxes equal in magnitude?

(iv) What is the net solar flux at the level determined in (iii) in terms of the direct solar flux at $\tau = 0$? Is the net flux upward or downward?

4.7 Invariant imbedding.** *Invariant imbedding* is another numerical approach to radiative transfer problems for a stratified atmosphere. It uses four, simultaneous, integro-differential equations involving slab reflection functions and slab transmission functions. As for doubling-and-adding, these equations are derived from conservation equations for finite slabs, rather than from the equation of transfer.

The reflection function, $R(\tau; \mu, -\mu_0)$, for a slab of optical thickness, τ, is defined from the integral relation,

$$I_r(\mu, \tau) = 2 \int_0^1 R(\tau; \mu, \mu_0) I(-\mu_0) \mu_0 \, d\mu_0.$$

For simplicity, all quantities are azimuthal averages, although the method is not restricted to such. $I_r(\mu)$ is an upward, reflected radiance; without a suffix the radiance is incident, and downward. Note that μ is defined as positive definite, equation (4.36). I_r will change its role and be an incident radiance for a slab lying above. A complete expression for the radiance requires the addition of a transmitted term from below, but this need not concern us here.

To a layer of optical thickness τ add an infinitesimally thin layer, of optical thickness, $\Delta\tau$, see Figure 4.13. Because the added layer is infinitesimally thin, only single scattering can occur in the layer. Five possible events are illustrated in the figure.

(i) Show that the reflected radiation from $\tau + \Delta\tau$ is,

$$I_r(\mu, \tau + \Delta\tau) = I_r(\mu, \tau) + 2\Delta\tau \int_0^1 I(-\mu_0) \frac{\partial R(\tau; \mu, -\mu_0)}{\partial \tau} \mu_0 \, d\mu_0.$$

(ii) Evaluate the five different terms that contribute to $I_r(\mu, \tau + \Delta\tau)$ and hence determine five terms in $\frac{\partial R(\tau; \mu, -\mu_0)}{\partial \tau}$. Taken together, these form one of the integro-differential equations for the method of invariant imbedding. Use the result of Equation (4.44) with P independent of ϕ.

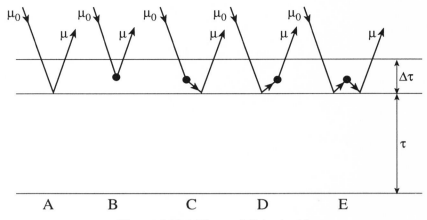

Figure 4.13 Adding an infinitesimal layer.

4.8 **Doubling and adding.**** By combining two layers, a and b, into a single layer c, obtain equation (4.42) from (4.39) and (4.41).

4.9 **Radiation to space.** The two most important bands of carbon dioxide in the thermal spectrum have absorption maxima at 667 cm^{-1} and 2349 cm^{-1}. The sums of the strengths of all lines in the two bands (the *band strengths*) are,

Band, cm^{-1}	$\sum_{\text{band}} S_{\text{line}}$, m^{2}Hz
667	2.83×10^{-13}
2349	3.17×10^{-12}

(i) The line strength, equation (3.7), involves a frequency unit. In the above table this unit is the hertz. Calculate the Planck functions at the band centers in these units for a temperature of 220°C. The bands are sufficiently narrow that these Planck functions may be applied to the entire bands.

(ii) Obtain an expression for the maximum potential rate of change of temperature $\left(\frac{\partial T}{\partial t} = \frac{\dot{q}}{c_p}\right)$ in terms of the band strength and the molar mixing ratio for the cooling-to-space approximation. $E_2(x) \to 1$, as $x \to 0$.

(iii) Calculate the maximum value of $\frac{\partial T}{\partial t}$ for each band individually, for a carbon dioxide molar mixing ratio of 3.45×10^{-4}.

CHAPTER 5

CONSTRAINTS ON THE THERMAL STRUCTURE

The previous three chapters have provided some of the tools needed to understand aspects of the vertical and horizontal thermal structure of the atmosphere. We examine some of the important issues in terms of simplified models.

Radiative equilibrium is the simplest assumption of any significance that can be made. Calculations based upon radiative equilibrium show resemblances to the observed atmosphere, even when the absorption coefficient is assumed to be independent of wavelength (grey absorption).

A decrease of temperature with height in the lower atmosphere and a trend towards an isothermal "lower stratosphere" are features of all radiative-equilibrium models. They also all exhibit unstable regions adjacent to the ground, which break down and form the convective troposphere. Radiative-convective models, in which an approximate representation of a convective layer is added to a radiative-equilibrium model, have been popular tools for climate research since the beginning of the century. A number of different assumptions may be made about the convective region. One is that it has a constant lapse rate of 6.5 K km^{-1}. We also discuss a tropospheric model that is based upon thermal transport by cumulus convection, an important process in the tropics.

Simple models may also be used to examine the meridional structure of atmospheric temperature. A strong constraint is the spherical shape of the planet, which leads to large insolation in the tropics and small insolation in polar regions. Atmospheric motions transfer energy from equator to poles and act to decrease the meridional temperature gradient. An order-of-magnitude calculation for the three inner planets—Venus, Earth, and Mars—shows that their different meridional temperature structures reflect an approximate balance between radiative and dynamical transports.

An important factor in the meridional structure of the planet is the presence or absence of ice and snow cover, a phenomenon closely connected to the surface temperature. This may be studied with surface, energy-balance models. An ice-covered surface reflects more solar radiation and has a lower equilibrium temperature than one that is ice-free. This effect is large enough that ice can, in principle, be thermally stable in tropical regions. Both ice-free and ice-covered solutions can exist for most latitudes.

Some essential features of the sensitivity of climate to external forcing may be studied with a simplified radiative-convective model. The essence of this model is that the atmospheric temperature hinges about the temperature of the emission layer, the layer from which thermal radiation is emitted to space. This highly simplified climate model is surprisingly effective and allows the investigator to calculate such matters as the effect of doubling the amount of carbon dioxide and the positive feedback resulting from the dependence of water-vapor density on atmospheric temperature by means of back-of-the-envelope calculations.

The effect of clouds on climate may also be studied with this simple model. The effect of clouds on surface temperature can be very large, with either heating or cooling, depending upon whether the clouds are low or high, respectively.

In a final section we discuss the inverse radiative transfer problem, that is, the retrieval of data from radiances measured by satellites in earth orbit. The atmospheric temperature may be retrieved with good accuracy, but with poor vertical resolution.

5.1 Vertical structure

5.1.1 Radiative equilibrium (approximate theory)

Thermal and solar radiation are the only significant external energy sources for the combined atmosphere-ocean system. These two radiation streams could, in principle, come into a state of radiative equilibrium, but this does not happen because of heat transfer by atmospheric motions. These motions are forced by the distribution of radiative sources and sinks of energy, so that radiative equilibrium on a spherical planet is, in principle, impossible. Nevertheless, instructive lessons may be learned from a study of this idealized state.

There are many numerical techniques by which the radiative equations may be iterated to find a temperature distribution that has zero net heating at every level. In Figures 5.2 and 5.3 we shall come to examples of such calculations, but more insight is provided by discussing a highly simplified, analytical, *semi-grey, two-stream model*. This model has serious limitations, and judgment is needed to appreciate the insights while, at the same time, understanding the limitations.

These are the characteristics of the model. The lowest order of the method of moments, equation (4.63), is employed (this is the two-stream part of the approximation). Inspired by the small overlap of the solar and terrestrial radiances in Figure 1.3, the two are treated independently, each with different wavelength-independent optical properties (the word *grey* implies wavelength independence, so this is the semi-grey part of the approximation). Other assumptions are not essential to the treatment. On the basis of the appearance of Figure 1.3, we shall assume that the atmosphere is transparent in the solar spectrum, but that scattering and absorption may take place at or near the surface. Finally, because of the long thermal wavelengths, we assume that there is no scattering in the thermal spectrum, $a_T = 0$.

With these assumptions we may integrate radiances and fluxes over either the solar spectrum or the thermal spectrum, and omit ν-suffixes. For radiative equilibrium in a stratified atmosphere, see §4.1.2,

$$\nabla \cdot \vec{F}(\text{rad}) = \frac{\partial F_z(\text{rad})}{\partial z} = 0 , \qquad (5.1)$$

or

$$F_z(\text{rad}) = F_T + F_S = \text{constant} = 0 . \qquad (5.2)$$

F_T and F_S are the vertical thermal and solar fluxes, respectively. Averaged over the entire planet and assuming a steady state, their sum must be zero at the upper limit to the atmosphere, which is why the constant in (5.2) is zero. According to our model assumptions, there is no absorption of solar radiation and F_S is constant; so, therefore, must be F_T. Because solar radiation is directed downwards, F_S is negative and F_T is positive.

The solar flux, F_S, sets the energy scale for climate problems. In the following sections we shall quote from calculations that use many different values of F_S, and we pause to consider what different values imply. Our discussion concerns climatic averages, which we assume to be in a steady state, and to refer either to the entire globe or to important latitude zones, such as the tropics. According to equation (4.64), the vertical flux of the direct beam of solar radiation is

$$F_\odot = \xi_\odot f , \qquad (5.3)$$

where f is the solar irradiance. We may use (5.3) for climate averages by employing an appropriate average values for the cosine of the solar zenith angle, $\bar{\xi}_\odot$. If we require the total energy falling on the entire globe to be conserved, this directional cosine must be minus the ratio of the cross-sectional area to the surface area of a sphere,

$$\bar{\xi}_\odot = -\frac{1}{4} . \qquad (5.4)$$

Similarly, for an average over tropical regions, it is minus the ratio of the diameter to the circumference of a sphere,

$$\bar{\xi}_\odot = -\frac{1}{\pi} . \qquad (5.5)$$

For the 1980 value of $f_0 = 1373$ W m^{-2}, see Appendix D, (5.4) and (5.5) lead to global and tropical averages of the direct solar fluxes of 343 and 436 W m^{-2}, respectively.

We must now take account of solar radiation scattered from the surface or from low clouds. This is a very variable quantity. Scattering from the ocean surface is about 5% of the incident solar radiation; from a thick, low cloud it may be as much as 80%. We introduce the *planetary albedo*, A, or mean reflectivity, with a typical range of $0.05 < A < 0.8$. It is commonly assumed that the global albedo is 0.12 for the hypothetical case of no clouds; it is 0.31 for an average cloud cover of $\sim 50\%$. The radiation scattered back into space is $AF_\odot$, and the solar net flux is, therefore,

$$F_S = \bar{\xi}_\odot f (1 - A) . \tag{5.6}$$

Thus, the global average solar flux for the canonical value of $A = 0.31$ is 236.8 W m^{-2}. Different values of the solar flux that are used by different authors reflect their assumptions about latitude zones and cloudiness.

The zenith angle determines not only the relationship between flux and solar irradiance, but also the length of the absorption path to be used when calculating the irradiance at levels inside the atmosphere. Our model contains no solar absorption in the atmosphere, so the distinction is unimportant, but if absorption were to be included we would be faced with an awkward aspect of nonlinear averaging, namely, that different values of $\bar{\xi}_\odot$ are required for the average absorption path. The atmosphere does not react strongly to day-night changes, and the values of $\bar{\xi}_\odot$ that we have discussed so far are appropriate for averages over day and night. However, absorption only occurs during the day, and the average absorption path is better calculated on the basis of the average solar zenith angle during the day. The reader is left to consider this. For a global average, we might use $\bar{\xi}_\odot = -\frac{1}{4}$ to calculate the solar flux and $\bar{\xi}_\odot = -\frac{1}{2}$ (a zenith angle of $60°$) for a simultaneous irradiance calculation. An example of this "inconsistency" can be found in Figure 5.2.

We now return to the approximate radiative equilibrium calculation. The equation governing the thermal flux is obtained from equation (4.63) with $a_T = 0$, and $F_T = $ constant:

$$4\pi \frac{dB}{d\tau} = 3F_T , \tag{5.7}$$

or

$$B(\tau) - B(0) = \frac{3F_T \tau}{4\pi} . \tag{5.8}$$

The suffix T has been omitted from the τs, but should be understood.

The boundary conditions were discussed, in principle, in §4.3.2. From equation (4.58) with constant F_T and $a_T = 0$,

$$\bar{I} = B . \tag{5.9}$$

From equation (4.67), the lower boundary condition at $\tau = \tau_1$ is

$$B_g - B(\tau_1) = \frac{F_T}{2\pi}, \tag{5.10}$$

where B_g is the emission of the surface.

Similarly, the boundary condition at $\tau = 0$ may be shown to be

$$B(0) = \frac{F_T}{2\pi}. \tag{5.11}$$

The second boundary condition determines the temperature structure in the atmosphere, while the first determines the temperature of the ground. The complete solution is

$$B(\tau) = \frac{\sigma}{\pi} T^4(\tau) = \frac{F_T}{2\pi}\left(1 + \frac{3\tau}{2}\right), \tag{5.12}$$

$$B_g = \frac{\sigma}{\pi} T_g^4 = \frac{F_T}{2\pi}\left(2 + \frac{3\tau_1}{2}\right). \tag{5.13}$$

An unfamiliar feature of these equations is that there are discontinuities in the thermal source function and, therefore, in the temperature, between the atmosphere and both boundaries. The solution is shown graphically by the full lines in Figure 5.1. The height scale in Figure 5.1 is chosen to simulate approximately the behavior of water vapor, the principal absorber in the lower atmosphere. The optical depth is taken to follow an exponential relation:

$$\tau(z) = \tau_1 \exp\left(-\frac{z}{H}\right), \tag{5.14}$$

with $H = 2$ km. Equation (5.14) represents approximately the observed distribution with height of the total water column above a level z in the troposphere (see the data in Appendix E).

5.1.2 Convection in the lower atmosphere

Below 20 km there are points of similarity between the equilibrium profile for $\tau_1 = 1$ in Figure 5.1 and the U.S. Standard Atmosphere in Figure 1.2. For both, temperature decreases with height below 20 km and tends asymptotically to a limit. In radiative equilibrium models this limit is called the *skin temperature* and is equal to 206.6 K in Figure 5.1. Details of the lower atmosphere are not reproduced well by the model: In particular, the temperature discontinuity at the surface (22.8 K in the case of $\tau_1 = 1$) is not observed in practice. This is hardly unexpected because the surface layers in Figure 5.1 are statically unstable, according to the criteria of §2.2.5, and will break down into free convection.

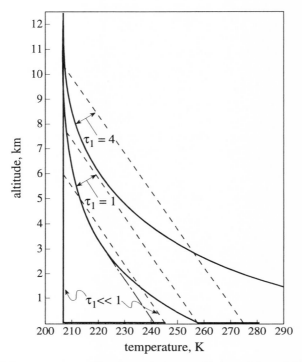

Figure 5.1 Temperature profiles for radiative equilibrium. The solid lines follow the solution (5.11), (5.12), and (5.13). Heavy lines indicate the surface temperature discontinuities. Broken lines indicate a lapse rate of 6.5 K km^{-1}. The calculations are based on a flux, $F_T = 206.3$ W m^{-2}. Some numerical values follow.

	$T(z=0)$, K	T_g, K	$-\left(\frac{\partial T}{\partial z}\right)_0$, K km^{-1}	T'_g, K	z_t, km
$\ll 1$	206.6	245.7	0.0	245.7	6.02
1.0	259.7	282.5	19.5	258.2	7.70
4.0	335.8	347.2	36.0	274.7	10.20

Definitions:

$T(z=0)$ = radiative equilibrium temperature at $z = 0$;
T_g = radiative equilibrium temperature of the surface;
T'_g = surface temperature for a convective troposphere;
$-\left(\frac{\partial T}{\partial z}\right)_0$ = lapse rate at $z = 0$;
z_t = height of the "tropopause."

There are two sources of instability in the surface layers. First, all profiles have an unstable surface discontinuity. Second, for $\tau_1 = 1$ and 4, there is a substantial region of the lower atmosphere in which $\Gamma > \Gamma_{ad}$. These configurations are unstable and must be replaced by profiles that takes into account entropy transport by free convection. Common assumptions are that the surface temperature discontinuity will be destroyed, and the lapse rate will be replaced by the observed mean lapse rate in the atmosphere, Γ_{obs}, which is usually chosen to be 6.5° K km^{-1}. The justification for this assumption is that for climate means (see Figures B.2 and B.3) Γ_{obs} does not vary greatly with latitude or with season.

An alternative procedure, which is partially supported by laboratory and theoretical studies, is to set $\Gamma = \Gamma_{ad}$, using moist or dry adiabats, as appropriate. Free convection carries entropy upwards with great efficiency, which will reduce Γ if $\Gamma > \Gamma_{ad}$. The result is to force Γ to be very close to Γ_{ad}; laboratory and atmospheric studies suggest that for distances more than a few meters from the ground, the two lapse rates will be indistinguishable.

We first consider *6.5 K km^{-1} adjustment models*. Referring to Figure 5.1 for $\tau_1 = 1$ as an example, we anticipate that the radiative-equilibrium region below 3.2 km should be replaced by a convective region with the chain dot-dash profile (the region appears to be smaller in the figure). But the convective region cannot stop here, for the following reasons.

Above the convective region the temperature profile, (5.12), will be unchanged because it was derived from the upper boundary condition, (5.11), which is unchanged. In this region the net thermal flux is also unchanged. In addition, the downward flux component F^- (see equation 4.35) is also unchanged because it is a function only of temperatures above the level concerned and these are unchanged. The upward flux component, F^+, on the other hand, is modified by the changes in the troposphere. In fact, because the convection, as represented by the chain dot-dash line in Figure 5.1, gives lower atmospheric temperatures *everywhere*, it follows that F^+ must be smaller than its radiative-equilibrium value *everywhere* in the radiative region. Consequently the net flux in the radiative region cannot be F_T, and the solution is not satisfactory. It may be made consistent only if the convective region is allowed to rise until F^+ at the transition between the two regions is equal to its required value. This is a soluble numerical problem with a unique solution.

Numerical integrations show that for $\tau_1 = 1$ the convective region must rise to 7.70 km before all of the requirements of the problem are satisfied. Surface temperatures are decreased by convection from T_g to T'_g; some numbers are given in Figure 5.1. Under all circumstances, however, the surface temperature is higher than it would have been in the absence of an atmosphere, which is 245.7 K for these model conditions. This difference is

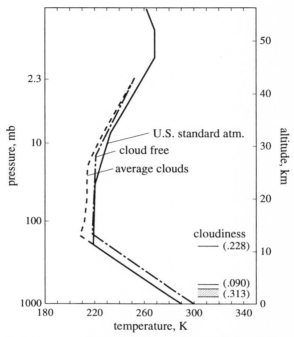

Figure 5.2 Radiative-convective calculations using detailed radiative computer codes. The convective lapse rate is assumed to be 6.5 K km^{-1}. The chain dash-dot line is calculated without clouds, with a solar flux equal to 299 W m^{-2}. The broken curve is calculated for the cloud conditions at the lower right of the figure. The clouds are treated as partially reflecting layers; the cloud amounts are shown in parentheses; the average reflectivities are taken from observation. The net solar flux for the cloudy atmosphere is 228 W m^{-2}. The solar zenith angle is 60°. The full line is the U.S. Standard Atmosphere, see Table B.1.

the atmospheric *greenhouse effect*. The name is more evocative than precise because the elevated temperature in a domestic greenhouse is primarily the consequence of the lack of wind and evaporation, rather than the radiative properties of the glass. For the earth as a whole the atmospheric greenhouse is about 30 K, corresponding roughly to the case $\tau_1 = 4$ in Figure 5.1.

The calculations shown in Figure 5.2 are also based upon a 6.5 K km^{-1} convective troposphere, but the radiative calculation is much more elaborate. A detailed numerical radiation code based on realistic optical properties is solved by iterative numerical methods. In the lowest 15 km, the agreement between the approximate calculations with $\tau_1 = 4$ and the cloudy profile in Figure 5.2 (these have similar solar fluxes) is remarkably close. Above 15 km, there are large differences between the models because of solar absorption by ozone that is included in the more precise model but not in the grey model. Typical solar heating rates are illustrated

CONSTRAINTS ON THE THERMAL STRUCTURE

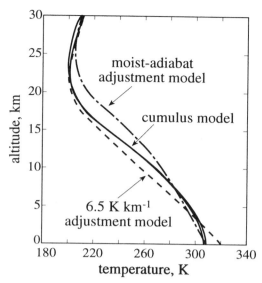

Figure 5.3 Radiative-convective profiles for three different tropospheric models. The radiative treatment is the same for all three models. The calculations are for average tropical conditions with a solar flux of 353 W m^{-2}. The three tropospheric models are discussed in the text.

by the data presented in Figure 2.4.

Figure 5.3 compares three tropospheric models. The 6.5 K km^{-1} adjustment model has been discussed. The *moist-adiabatic adjustment model* replaces the constant 6.5 K km^{-1} lapse rate by the moist adiabatic lapse rate. Below 5 km in the tropics, the moist adiabat is less than 6.5 K km^{-1} (see Table 2.1). The *cumulus convection model* will be discussed in the next section; it is typical of a class of one-dimensional models based on the behavior of tropical cumulus clouds, which are the principal means of vertical entropy transport in the tropics.

5.1.3 A cumulus convection model

Some details of a simplified cumulus convection model are given in Figure 5.4. The updraft in the core (subscript c) is fast and may be treated as adiabatic (see §2.2.3); it also has accelerations that are too great for the core to be treated as hydrostatic (see equation F.12). The environment (subscript e), on the other hand, moves slowly and may be treated as hydrostatic, and the nonadiabatic nature of the flow in the environment is an essential feature of the model. There is no horizontal discontinuity of pressure at the core-environment boundary; such discontinuities dissi-

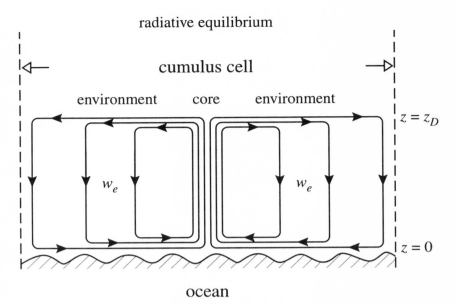

Figure 5.4 Schematic representation of a tropical cumulus cloud. The *core* of the cumulus cell is assumed to be of negligible lateral extent. *Detrainment* and *entrainment* are horizontal motions that occur at the top and bottom of the cell, respectively, and are assumed to have negligible vertical extent. The model consists principally of the *environment* in which there is a steady downward velocity, w_e. Environmental parameters are assumed to be functions of height only. The core acts to create an adiabatic path from the entrainment region at the surface to the detrainment level at $z = z_D$. Consistent with a tropical model, the lower boundary is assumed to be an ocean.

pate with the speed of sound and may be neglected in most atmospheric problems. This means that the pressure-height relationship created by hydrostatic equilibrium in the environment is impressed upon the core. We now assume that the core is very small compared with the environment, so that the atmospheric model is, in essence, the environment alone; the function of the core is restricted to providing a thermal connection between the top and bottom of the convection system.

Temperature profiles are illustrated in Figure 5.5. At the *detrainment level*, z_D, the core and environment are connected by a horizontal flow and the temperature, T_D, is the same in both. Above this level there are no motions, and the atmosphere is in radiative equilibrium; below, the temperature of core and environment must be determined independently from appropriate thermodynamic models. We shall now start by assuming the atmospheric temperature profile, $T_e(z)$, and the detrainment level, z_D, and iterate both until the thermodynamic equations are satisfied, with appropriate boundary conditions at the detrainment level and at the surface.

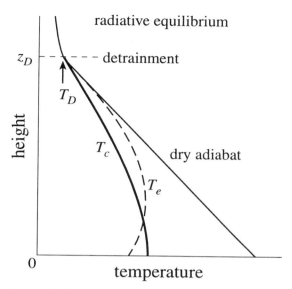

Figure 5.5 Temperature structure for cumulus convection.

Procedurally this is the same as for the radiative-convective models in the previous section and is equally straightforward for computers to handle, but the procedure is much more complicated to describe.

First note that, given $T_e(z)$ and z_D, the thermodynamic state of the core is fully defined, including the surface temperature, $T_c(0)$. This follows because the core is saturated at the surface, where it is in contact with water, and at all levels above because the air cools in the adiabatic updraft. Because the flow in the core is adiabatic, the equivalent potential temperature is the same at all levels. It is known at the detrainment level because the pressure there may be calculated from hydrostatic equilibrium in the environment (given the surface pressure), the temperature has been assumed, and the air is saturated. It follows that the pressure, the temperature, and the humidity in the core may be calculated at all levels below z_D.

Now consider the environment. The thermodynamic equation is (2.17). We neglect $\dot{\phi}_{\mathrm{irr}}$ and horizontal motions, assume a steady state, and make use of (2.39):

$$\frac{ds}{dt} = w_e \frac{\partial s}{\partial z} = \frac{w_e c_{p,a}}{T_e}\left(\frac{dT_e}{dz} + \Gamma_{\mathrm{dry}}\right) = \dot{q}\,. \qquad (5.15)$$

To calculate $\dot{q}$ we require the water-vapor mixing ratio in the environment (the densities of other gases are assumed to be known). This is constant in the sinking air of the environment and equal to its value at z_D, which is known. In addition we need to know the temperature at all levels (which we have assumed) and the temperature of the surface. Let us first assume that the surface temperature is equal to the temperature at the bottom of the core, which we know. We may now calculate $\dot{q}$ at all levels and, from (5.15) and the given temperature at z_D, we may calculate $T_e(z)$. In general, this solution will not correspond to the temperature profile assumed initially, but there will be a unique temperature profile that is self-consistent, which may be determined iteratively.

At this point we have obtained self-consistent solutions in the core and the environment given the four parameters, T_D, z_D, T_g (assumed equal to T_c at $z = 0$), and w_e. The remainder of the calculation consists in iterating these parameters until boundary conditions and stability conditions are satisfied.

Above z_D the temperature profile is determined from a radiative equilibrium calculation. There is no reason why this should lead to temperature continuity at z_D. However, temperature continuity is the only stable configuration, as it was for the 6.5 K km^{-1} adjustment model. We may change T_D until we find a self-consistent solution without a discontinuity.

We have made no effort to constrain the outgoing thermal flux at the top of the convective region, but this is an externally imposed parameter in climate models. We may calculate the outgoing flux and iterate z_D (and, of course, all other parameters) until the required flux is obtained.

For the remainder of this section we assume w_e to be a given external parameter and we calculate T_g from a surface heat flux condition. Other approaches are possible, but this illustrates the problems involved. We consider the ocean, average over an area corresponding to a single cumulus cell, and assume a steady state. Only vertical fluxes through the ocean atmosphere boundary are considered, and we neglect internal dissipation. With these assumptions, equation (2.66), applied to the ocean, yields

$$\overline{F}_z(e + \pi, z = 0) + \overline{F}_z(\text{rad}, z = 0) = 0 \ . \tag{5.16}$$

The overbar indicates an average over the cumulus cell. According to equation (2.68) (neglecting the third term on the right) and using the definition (F.21), the average flux per unit area of total potential energy is

$$\overline{F}_z(e + \pi, z = 0) = w\rho(c_v T + lm_v) \ . \tag{5.17}$$

w is the vertical velocity above the entrainment layer, and m_v is the water-vapor mixing ratio. We now assume that there are only two surface temperatures, one for the core $T_c(0)$, and one for the environment, $T_e(0)$, and that m_v has its saturated value in the core, but may be neglected in the environment. The integral of $w\rho$ over the cell is the net flux of the atmosphere itself at the lower surface and is zero. Averaging core and environment together, we have

$$\overline{F}_z(e+\pi, z=o) = w_e\rho\{c_v[T_e(0) - T_c(0)] - lm_v[T_c(0)]\} \ . \tag{5.18}$$

Finally, for the radiative flux, we have

$$\overline{F}_z(\text{rad}, z=0) = \sigma T_g^4 - F^-(\text{rad}, z=0) \ . \tag{5.19}$$

Equations (5.16), (5.18), and (5.19) together yield a relation between T_g and $T_c(0)$, enabling us to eliminate the arbitrary assumption that they are equal. The problem is now closed, given the environmental velocity, w_e.

5.1.4 The Chapman layer

In this section we introduce three concepts that are invaluable for discussing radiative constraints on atmospheric structure: the emission temperature, the emission level, and the Chapman layer.

The *emission temperature*, T_e, is defined by equating the outgoing flux from the atmosphere to that from a black surface,

$$F_T(\tau = 0) = \pi B_e = \sigma T_e^4 \ . \tag{5.20}$$

The *emission level*, z_e (optical depth τ_e), is the level whose temperature is T_e. From equation (5.12),

$$B_e = B(\tau_e) = \frac{F_T}{2\pi}\left(1 + \frac{3\tau_e}{2}\right) = \frac{F_T}{\pi} \ , \tag{5.21}$$

so that

$$\tau_e = \frac{2}{3} \ . \tag{5.22}$$

The emission level is more than an abstract idea. Contributions to the radiative flux to space exhibit a layered structure, the *Chapman layer*, which peaks close to the emission level. To a first degree of approximation the atmosphere may be regarded as radiating to space from an isothermal layer at $\tau_e = \frac{2}{3}$, whose temperature is T_e.

To show this we first discuss the equations governing the radiance and then give an approximate relationship between radiance and flux. Start from equation (4.31) with, for the sake of convenience, $\tau_1 \gg 1$. Changing variables from t to z and evaluating the radiance at $\tau = 0$, we have

$$I_\nu^+(0, \xi) = \int_0^\infty B_\nu(z) h_\nu(z, \xi) \, dz \ , \tag{5.23}$$

where the *kernel function* is

$$h_\nu(z,\xi) = -\frac{1}{\xi}\frac{d\tau_\nu}{dz}\exp\left[-\frac{\tau_\nu(z)}{\xi}\right] . \tag{5.24}$$

Equation (5.24) is normalized:

$$\int_0^\infty h_\nu(z,\xi)\,dz = 1 . \tag{5.25}$$

Let $\tau_\nu(z)$ follow an exponential distribution with height, as expressed by equation (5.14). This assumption was briefly discussed for water vapor; it is also a reasonable assumption for mixed gases with H equal to the scale height of air. With this assumption,

$$h_\nu(z,\xi) = \frac{1}{H\xi}\tau_\nu(z)\exp\left[-\frac{\tau_\nu(z)}{\xi}\right] . \tag{5.26}$$

The kernel function, $h_\nu(z,\xi)$, has a single maximum at $\tau_\nu(z) = \xi$. This will correspond to the emission level if $\xi = \frac{2}{3}$.

Equation (5.26) is for monochromatic radiation, and it will be used in §5.4. Up to this point, however, the discussion has been restricted to grey absorption and to spectrally integrated radiances and fluxes. For this circumstance, the equations are also valid, but with the ν-suffixes omitted. In order to complete our discussion we must now show that the behavior of the flux is approximately the same as the behavior of the radiance for the zenith angle corresponding to $\xi = \frac{2}{3}$.

From the definition of flux, equation (4.5), and the geometry of Figure 4.3, we may use the mean value theorem to write

$$F^+(z) = I^+(z,\overline{\xi})\int_0^1 2\pi\xi\,d\xi = \pi I^+(z,\overline{\xi}) , \tag{5.27}$$

where $\overline{\xi}$ lies between 0 and 1. Many investigators have suggested values of $\overline{\xi}$. The value depends upon circumstances of the calculation, but all that have been proposed are close to $\frac{2}{3}$. For all practical purposes, the maximum of the kernel function for the flux to space is at the emission level.

The distribution function may be written in the form

$$\frac{h(z)}{h(\max)} = \frac{\tau(z)}{\tau(\max)}\exp\left[1 - \frac{\tau(z)}{\tau(\max)}\right] , \tag{5.28}$$

where

$$h(\max) = \frac{1}{eH} , \tag{5.29}$$

$$\frac{\tau(z)}{\tau(\max)} = \exp\left[\frac{z - z(\max)}{H}\right] . \tag{5.30}$$

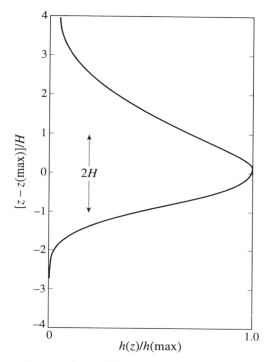

Figure 5.6 The Chapman layer. The figure has been constructed for $\tau_1 \gg 1$, but the only modification required for a finite atmosphere is to truncate the layer at the value of $\frac{z-z(\max)}{H}$ corresponding to the surface. That part of the distribution that lies above the surface is unchanged.

Equation (5.28) is plotted in Figure 5.6.

5.2 Meridional structure

5.2.1 Emission temperatures

The emission temperature, as defined by equation (5.20), is based upon the outgoing flux of thermal radiation. It can, however, also be based upon the absolute value of the incoming flux of solar radiation. For radiative equilibrium the two fluxes are equal and opposite in sign, and the resulting temperatures, $T_{e,T}$ and $T_{e,S}$, are equal; but with horizontal dynamical transports in the atmosphere, the two temperatures differ. Table 5.1 shows some data.

The ratio $x = \frac{\Delta T_{e,T}}{\Delta T_{e,S}}$ is a measure of the importance of horizontal transports. If $x = 1$, the thermal and solar fluxes balance and there is no

Table 5.1 Emission temperatures calculated for thermal and solar fluxes

Region	F_S W m^{-2}	$T_{e,S}$ K	$\Delta T_{e,S}$ K	F_T W m^{-2}	$T_{e,T}$ K	$\Delta T_{e,T}$ K
0-30°	301	270		250	258	
globe	236	254	35	236	254	8
30-60°	172	135		223	250	

$\Delta T_{e,T}$ and $\Delta T_{e,S}$ are temperature differences between tropical and extratropical regions. The data in the first and third rows are taken from observations. The second row is theoretical, based upon a global albedo of 0.31.

horizontal, dynamical transport. $x \ll 1$, on the other hand, means that the temperature is almost the same all over the planet and that the horizontal flow is effectively adiabatic. The question that this raises is similar to that discussed in §2.2.3, except that the heating term, $\left|\frac{\dot{q}}{c_p}\right|$, must be compared with a horizontal transport term. For meridional transport, the appropriate term is $\left|v\frac{\partial \theta}{\partial y}\right|$. To order of magnitude we may approximate[1]: $\dot{q} \sim \frac{\sigma T_e^4}{M}$, where M is the mass per unit area of an atmospheric column; $v \sim V$, where V is the observed magnitude of meridional wind speeds; and $\frac{\partial \theta}{\partial y} \sim \frac{T_e}{R}$, where R is the radius of the planet. The last approximation carries with it a number of debatable assumptions about the meridional extent of the circulation and the meridional temperature contrast, but it serves to make a qualitative point. With these approximations,

$$\frac{\dot{q}}{c_p} \bigg/ v\frac{\partial \theta}{\partial y} \sim \frac{\sigma R T_e^3}{V M c_p}. \qquad (5.31)$$

From this discussion we may anticipate that the ratio (5.31) and the ratio x are related; when one is small so should be the other, and vice versa.

Data for the inner planets are given in Table 5.2. Venus has a very small meridional temperature contrast, while for Mars the contrast is greater than for Earth. The temperature contrasts for the three planets and the parameter defined in equation (5.31) are compared in the final two columns. They agree as to order, indicating that our ideas are self-consistent. It should be noted that the strongest variable in equation (5.31) is the mass per unit area. A thin atmosphere is likely to be closer to radiative equilibrium than is a dense atmosphere.

---

[1] The reader may be unaccustomed to such approximate arguments. They are, however, an essential part of atmospheric dynamics.

CONSTRAINTS ON THE THERMAL STRUCTURE

Table 5.2 Data for the inner planets

	Global T_e, K	V, m s^{-1}	M, kg m^{-2}	x	$\frac{\sigma R T_e^3}{V M c_p}$
Venus	244	0.5	1.1×10^6	< 0.03	~ 0.001
Earth	251	0.5	1.0×10^4	0.23	0.1
Mars	216	2	1.6×10^2	0.9	1.5

5.2.2 Energy balance climate models

Energy balance models occupy an important place in the hierarchy of climate models. Despite their simplicity, they exhibit interesting behaviors that often persist in general circulation models.

The surface of the planet is divided into strips contained between two close latitude circles. The energy balance per unit area, averaged over a strip, may be expressed by a relationship,

$$C \frac{\partial T(x)}{\partial t} = \text{solar term} + \text{thermal term} + \text{dynamical term} . \tag{5.32}$$

C is a heat capacity, and x is the sine of the latitude.

The solar heating rate may be represented approximately as a function of latitude by

$$\text{solar term} = \frac{f}{4} s(x)[1 - A(x)] . \tag{5.33}$$

$A(x)$ is the surface albedo of a latitude strip; $s(x)$ is a normalized distribution function for the solar flux. A function that incorporates, approximately, all of the geometric factors involved is

$$s(x) = 1 - 0.241(3x^2 - 1) . \tag{5.34}$$

Thermal radiation cools the surface. On the basis of observed conditions, the thermal term has been approximated by the linear relation,

$$\text{thermal term} = -[211.1 + 1.55 T_g(x)] . \tag{5.35}$$

$T_g(x)$ is the surface temperature expressed in °C. Equation (5.35) is empirical and is suitable only for atmospheric conditions close to those that are currently observed. Applicability to other conditions, such as those of an ice age (one of the reasons for introducing energy balance models in the first place), is speculative.

The dynamical term has been expressed in a number of ways by different investigators. One example is a diffusion equation:

$$\text{dynamical term} = \frac{\partial}{\partial x}\left[(1-x^2)D\frac{\partial T_g}{\partial x}\right], \quad (5.36)$$

where D is a diffusion coefficient.

A feature of this class of model is to represent the climatic response to changes in the surface albedo from that of oceans or dry land when the temperature is high to the much higher albedo of snow and ice when the temperature is low. The changeover from one type of surface to the other is assumed to take place at a temperature equal to $-10°C$, and the albedos are taken to be

$$\text{ice or snow}: A_i = 0.6, T_g < -10°C, \quad (5.37)$$
$$\text{ocean or land surface}: A_s = 0.3, T_g > -10°C. \quad (5.38)$$

Again, equations (5.37) and (5.38) incorporate a great deal of empirical information and should only be used under conditions close to those that now occur.

We consider the steady state ($\frac{\partial}{\partial t} = 0$), with no dynamical term ($D = 0$), thus avoiding two very uncertain features of energy balance climate models. With these restrictions we may still discuss two interesting matters. First, the surface temperature as a function of latitude for the 1980 insolation ($f = 1373$ W m^{-2}) is,

$$T_g(x) = \frac{(1-A) \times 1370}{4 \times 1.55}[1 - 0.241(3x^2 - 1)] - \frac{211.1}{1.55}, \quad (5.39)$$

and, second, the relation between the latitude of the ice line (x_i where $T_g = -10°C$) and the irradiance,

$$f = \frac{(211.1 - 10 \times 1.55) \times 4}{(1-A)[1 - 0.241(3x_i^2 - 1)]}. \quad (5.40)$$

Equations (5.39) and (5.40) are plotted in Figures 5.7 and 5.8, respectively. Both repay careful study because it is the only way to understand their rather strange behaviors. The most revealing approach is to follow the circuits indicated by the arrows. In Figure 5.7, start from the tropics ($x = 0$) under ice-free conditions, move to the pole, and then return to the tropics. In Figure 5.8 start from $f = 0$ with an ice-covered planet, increase the insolation to $f = 4000$ W m^{-2}, when the planet will be free of ice, and then reverse the events. Hysteresis is a major feature of both circuits, the reason for which may be illustrated by an example. Suppose that $\frac{fs(x)}{4} = 323.71$ W m^{-2}. From (5.35) we find $T_g = +10°C$ for $A = 0.3$

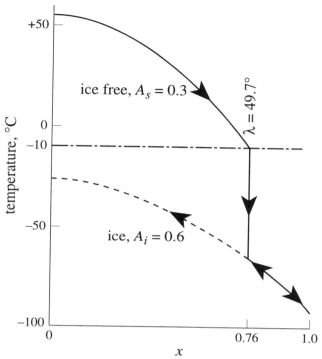

Figure 5.7 Temperature as a function of latitude. The calculations follow from (5.39), using the albedos (5.37) or (5.38), as appropriate to the surface temperature. The arrows follow an imaginary journey from equator to pole and back. The ice line can lie anywhere from the equator ($x = 0$) to latitude 49.7° ($x = 0.76$).

(ice-free condition), but $T_g = -52.7°C$ for $A = 0.7$ (ice cover). If the surface is ice free it stays ice free because $T_g > -10°C$; but if it is frozen, it stays frozen because $T_g < -10°C$. The previous history of the surface affects its current state. According to Figure 5.7, with the present insolation, all latitudes less than 49.7° have this ambiguity. In particular, ice at the equator appears to be a possible solution.

The hysteresis in Figure 5.8 has the same origin, but the context differs. Again we see that in the absence of dynamical transport the ice line for the present insolation may be at the equator or at 49.7°. However, also shown are the results for a small value of the transfer coefficient in equation (5.36). Hysteresis has disappeared. There is now only one curve, and this does not have ice at the equator for the present insolation. Instead, two solutions with ice lines between the equator and 49.7° have appeared. The solution close to the equator may be shown to be unstable.

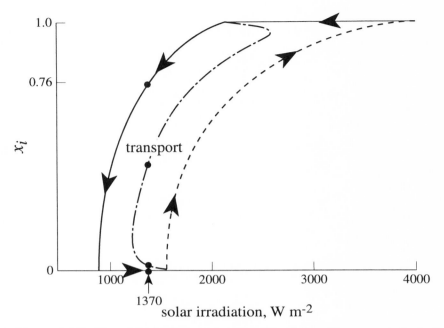

Figure 5.8 Position of the ice line as a function of insolation. The arrows indicate the result of increasing the insolation from very low to very high values, and subsequently decreasing it. The curve marked "transport" is a solution for a small value of D in equation (5.36). The filled circles are multiple solutions for the present insolation, both with and without a small dynamical transport.

5.3 Climate sensitivities

In this section we discuss the sensitivities of climate to forced changes. The usual approach is to calculate the climate, using a numerical model, with and without the forced change. While this procedure yields results, it yields little insight into mechanism. By way of contrast, we shall look at an exceptionally simple climate model in which the mechanisms are transparent, even if the results may be considered to be less reliable. The *sensitivities* of this simple model, that is to say, the responses of the model to forcings, differ little from those obtained from complex numerical models.

5.3.1 A simple greenhouse model

The model is based upon our discussion of the Chapman layer in §5.1.4. The strongest constraint upon the climate is the first law of thermodynamics, expressed through the numerical value of the emission temperature. We saw that this temperature can be assigned to the emission level, establishing

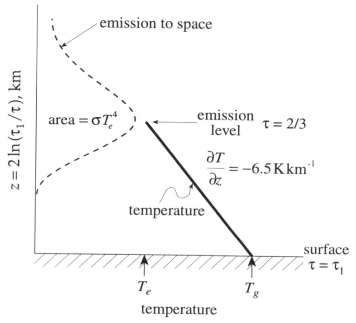

Figure 5.9 A simple greenhouse model.

one point on the atmospheric $T - \tau$ profile, about which the atmosphere must hinge. This situation is illustrated in Figure 5.9.

For a semi-grey model, the height of the emission level is, from (5.14) and (5.22),

$$z_e = H \ln \frac{3\tau_1}{2}, \tag{5.41}$$

where for water vapor $H = 2$ km.

If we accept that the tropospheric lapse rate is $\Gamma_{\rm obs} = 6.5°{\rm K\ km}^{-1}$, the temperature of the emission level and the temperature of the surface are connected by

$$T_g = T_e + \Gamma_{\rm obs} z_e = T_e + \Gamma_{\rm obs} H \ln \frac{3\tau_1}{2}. \tag{5.42}$$

There is a great deal of physical information in equation (5.42). T_e depends upon the solar irradiance and the planetary albedo, involving questions of surface ice and snow and the presence or absence of clouds. τ_1 varies with the amounts of absorbing gases in the atmosphere. H and Γ are determined by many processes, principally dynamical. With $F_s = -236.3$ W m^{-2}, $\Gamma_{\rm obs} H = 13$ K, and $\tau_1 = 8$ we find,

$$T_e = 254.1 \text{ K},$$

$$T_g = 286.4 \text{ K},$$
$$z_e = 5.0 \text{ km}.$$

The total optical depth of the atmosphere was chosen to be $\tau_1 = 8$ in order to give an acceptable value for T_g. Semi-grey absorption is an extreme approximation to the atmospheric line absorption spectrum, and a rational estimate of τ_1 is not possible. For this particular calculation the greenhouse effect is 32.3 K, which is close to that of the present atmosphere. We now discuss how T_e and T_g change as the parameters of the problem are changed.

5.3.2 Human influence on climate

A major preoccupation for climate research is to predict the consequences of anthropogenic changes in the concentrations of atmospheric absorbers. The canonical problem is to predict the consequences of doubling the amount of atmospheric CO_2, an event that could occur, according to some predictions, by the year 2100 AD.

The two most important absorbers of thermal radiation in the lower atmosphere are water vapor and carbon dioxide (see Figures 1.1 and 2.4). Pursuing the simple greenhouse model, we may write

$$\tau_1 = \tau_{1,c} + \tau_{1,w}, \tag{5.43}$$

where $\tau_{1,c}$ is the total optical depth of carbon dioxide and $\tau_{1,w}$ is the total optical depth for water vapor.

The sensitivity of the ground temperature to change in the carbon dioxide amount, if all other factors are held constant, is given by the expression,

$$\frac{\partial T_g}{\partial \ln \tau_{1,c}} = \Gamma_{\text{obs}} H \frac{\partial \ln \tau_1}{\partial \ln \tau_{1,c}} = \Gamma_{\text{obs}} H \frac{\tau_{1,c}}{\tau_1}. \tag{5.44}$$

In the lower atmosphere, CO_2 bands dominate about 12.5% of the thermal spectrum, and most of the remaining spectrum contains water vapor absorptions. We may estimate

$$\frac{\tau_{1,c}}{\tau_1} \approx 0.125,$$
$$\frac{\tau_{1,w}}{\tau_1} \approx 0.875.$$

From (5.44),
$$\frac{\partial T_g}{\partial \ln \tau_{1,c}} \approx +1.6 \text{ K},$$

or, if CO_2 doubles,
$$\Delta T_g \approx +1.1 \text{ K}.$$

5.3.3 Water-vapor feedback

The foregoing estimate of sensitivity lies close to the median of estimates made with more sophisticated models for the same circumstances. It refers to the change that would occur when carbon dioxide changes, provided that nothing else happens. Linkage between phenomena is, however, one of the most characteristic features of the atmosphere. A *water-vapor feedback* occurs because of the strong dependence of the saturated pressure of water vapor upon temperature (see Appendix E). Another feedback occurs through the effects of temperature and motions on cloud amounts and cloud types; this will be briefly discussed in the next section.

At normal ground temperatures the vapor pressure of water increases approximately by a factor of two if the temperature increases by 10 K. An increase in opacity due to carbon dioxide can lead to an increase in temperature, hence to an increase in water-vapor opacity, and to a positive feedback on the temperature.

Under the debatable assumption that the total atmospheric water is proportional to the density at the surface, we may write

$$\tau_{1,w} \approx \tau_{1,w,0} \exp\left(\frac{T_g - T_{g,0}}{10} \ln 2\right), \tag{5.45}$$

where the zero suffix indicates a reference state. Substituting (5.45) into (5.42) and (5.43) gives the sensitivity,

$$\frac{\partial T_g}{\partial \ln \tau_{1,c}} = \alpha \Gamma_{\text{obs}} H \frac{\tau_{1,c}}{\tau_1}, \tag{5.46}$$

where α is the *amplification factor*,

$$\alpha = \left(1 - \Gamma_{\text{obs}} H \frac{1}{\tau_1} \frac{\partial \tau_{1,w}}{\partial T_g}\right)^{-1},$$

$$= \left(1 - \Gamma_{\text{obs}} H \frac{\tau_{1,w}}{\tau_1} \frac{\ln 2}{10}\right)^{-1}. \tag{5.47}$$

With the numerical values introduced previously, $\alpha = 4.7$, and the change in T_g for a doubling of CO_2 becomes 5.2 K. This number is very sensitive to the parameters used in equation (5.47). As pointed out by Sir George Simpson in the 1920s, this is where a more realistic radiation model makes a difference. In particular, the 10 μm water-vapor window allows surface radiation to pass directly to space, which exerts a strong control over surface temperatures. This feature is absent from a grey model when $\tau_1 \gg 1$. Nevertheless, we use the grey model to explore qualitatively a further possibility, the *runaway greenhouse*.

With a slight increase in $\tau_{1,w}$ or, from (5.45), a slight increase in ground temperature, (5.47) could give an infinite amplification factor. The end point of this process, if it were to occur, would be the evaporation of the oceans and the creation of a massive, hot, water-vapor atmosphere. This alarming possibility is, fortunately, more remote than our simple theory suggests. For large water-vapor concentrations we come closer to reality by using the small absorption coefficient in most transparent part of the thermal spectrum, the 10 μm water-vapor window, as the coefficient of the grey atmosphere. According to our simple theory, we then require there to be an optical depth in the window of four or more before a runaway greenhouse can occur. For the continuum data near 1000 cm^{-1} in Figure 3.7, it has been estimated that this would require a saturated atmosphere with average surface temperatures in excess of 314 K. While this may appear to put a runaway greenhouse well outside practical concern, it is not irrelevant to point out that the runaway greenhouse is the accepted explanation for the 730 K surface temperature of Venus.

5.3.4 Clouds and climate

The nature and the amount of cloud in the atmosphere has a profound effect on the climate. At present, the average cloud cover of earth is $\sim$50%. A glance at a satellite photograph makes the point that clouds are ephemeral, variable, and hard to predict.

It is normal practice for radiation studies to make an imprecise division of clouds into *low clouds*, 4 km or less above the surface, and *high clouds*, typically cirrus, about 8 km above the surface. Low clouds are more reflective than are high clouds; reflectivities are commonly assumed to be 0.7 and 0.3, respectively, against an average ground reflectivity of 0.12.

If the cloud amount changes, the global albedo will change, and hence the emission temperature will alter. If nothing else happens then, according to the simple greenhouse model, this translates into a change in ground temperature equal to the change in emission temperature. But there is an equally important effect related to the heights of the clouds. Low cloud lies below the emission layer and, according to the simple greenhouse model, will affect the albedo but not the height of the emission layer; high cloud, on the other hand, will, in addition to changing the albedo, raise the emission level to the cloud tops, at 8 km. Referring to equation (5.42), we may either change T_e only (for low clouds), or we may change both T_e and z_e (for high clouds). These ideas may be combined with the simple greenhouse model to calculate surface temperatures with and without complete cover of high and low clouds, leading to the results in Table 5.3.

The difference between high and low clouds is dramatic; the former cool while the latter heat the surface, and the surface temperature changes, in both cases, are large. Predictions of cloud amount and optical properties to the precision required for climate research are matters of very great

Table 5.3 Ground temperatures with and without clouds

	No cloud			Complete cover			Difference
	A	z_e, km	T_g, K	A	z_e, km	T_g, K	ΔT_g, K
High clouds	0.12	5	335	0.3	8	359	+24
Low clouds	0.12	5	335	0.7	5	271	−63

The numbers for z_e and A are discussed in the text. The equations used are (5.6), (5.20) and (5.42).

difficulty.

5.4 Remote sensing

Remote sensing of atmospheric parameters is performed by analyzing outgoing radiance spectra, as recorded by a downward-viewing satellite spectrometer. The principles involved have much in common with our discussion of radiative constraints upon atmospheric vertical structure and are appropriately considered here.

Techniques of remote sensing are applicable to all atmospheric parameters, provided only that they leave some imprint on the spectrum of outgoing radiation, but the major application has been to temperature soundings, using CO_2 bands in the thermal spectrum. The mixing ratio of CO_2 is known at the relevant atmospheric levels, eliminating all variables but the temperature, provided that observing frequencies are chosen to avoid other absorbers.

Figure 5.10 shows an emission spectrum in the region 2100–2500 cm^{-1}. This region is dominated by a strong vibration band of $^{12}CO_2$ centered on 2349 cm^{-1}. The same vibrational transition, but for $^{13}CO_2$, is at 2283 cm^{-1}, and a narrow band of N_2O is at 2224 cm^{-1}. Rotation lines are not resolved because the spectrometer resolution of ~ 2 cm^{-1} is insufficient to do so. For the inversion under discussion, the 10 frequencies chosen are indicated by tags below the spectrum. Closer spaced frequencies are redundant, as will be apparent when kernel functions are discussed.

To understand the nature of the inversion process, we return to the discussion of the Chapman layer in §5.1.4. We saw that the emission to space had a layered structure, corresponding to the kernel function, and that it is approximately true that the emission to space corresponds to the source function (the Planck function) for the temperature at the maximum

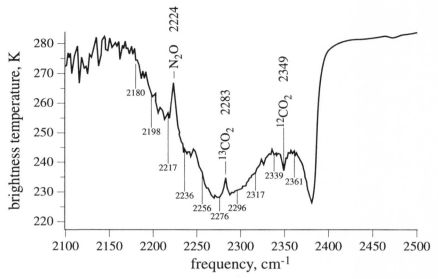

Figure 5.10 Brightness temperatures in the 2348 cm^{-1} CO_2 band. The relationship between the brightness temperature and the radiance is explained in the text. The three main bands in this spectral region are indicated by tags above the spectrum. The average resolution is 2 cm^{-1}, which smooths out all rotation lines, leaving only the envelope of the band. The frequencies used for analysis are shown by tags below the spectrum.

of the kernel function (equation 5.26). For nadir viewing by the satellite the kernel function is evaluated for $\xi = 1$, and the maximum of the kernel function then lies at $\tau_\nu = 1$. We now introduce the *brightness temperature* at frequency ν, $T_{b,\nu}$. This temperature is derived from the radiance by inverting the Planck function, equation (4.9),

$$T_{b,\nu} = \frac{h\nu}{k \ln\left(1 + \frac{2h\nu^3}{c^2 I_\nu}\right)}. \quad (5.48)$$

The brightness temperature is used to represent the outgoing radiance in Figure 5.10. It spans a range from 230 K to 280 K.

The discussion in §5.1.4 may now be interpreted to mean that the brightness temperatures shown in Figure 5.10 are approximately equal to the air temperatures at levels corresponding to $\tau_\nu = 1$. From the known spectral data and from the known vertical distribution of CO_2, this condition on the optical depth may be interpreted in terms of a height. Calculated kernel functions, as a function of height, are given in Figure 5.11. These are not precisely Chapman functions because the absorption coefficient is not, in reality, independent of height, nor are the data monochromatic, because the spectrometer has insufficient spectral resolution. Never-

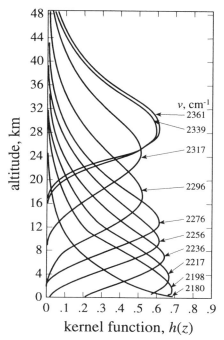

Figure 5.11 Kernel functions for the 2349 cm^{-1} CO_2 band. The numbers are the central frequencies of the regions chosen for analysis of the temperature inversion. They are also indicated in Figure 5.10.

theless, regardless of the complexity, the kernel function may be calculated and the height of its maximum may be obtained.

We see how, from the data in Figure 5.10, a number of atmospheric temperatures may be associated with a number of heights between 0 and 28 km, one pair for each frequency. For this particular data set 10 frequencies were chosen. The optimum number of frequencies depends on the degree of overlap of the kernel functions. If the kernel functions for two frequencies overlap strongly, the data are not independent; they may be used to increase the accuracy of the temperature determination, but not the vertical resolution. The 10 frequencies chosen were judged to be the maximum number with nonredundant information. The 10 independent (T, z) pairs give a first approximation to the temperature profile, $T_0(z)$.

A number of techniques have been developed that lead to the "best" temperature-height relationship for a given set of observed radiances. We shall look at a method proposed by Chahine, because it is particularly simple to describe. If $h_\nu(\max)$ and $T(\max)$ are the values of h and T at the maximum of the kernel function, then from the mean-value theorem applied to equation (5.23), the observed radiance may be written

$$I_\nu(0) = \int_0^\infty B_\nu[T(z)]h_\nu(z)\,dz \approx Ch_\nu(\max)B_\nu[T(\max)], \qquad (5.49)$$

where C is an unknown number. $h_\nu(z)$ is evaluated for zenith viewing, $\xi = 1$.

We now start with the first-order temperature profile that we have discussed, $T^0(z)$. For this profile we calculate an outgoing radiance,

$$I_\nu^0(0) = \int_0^\infty B_\nu[T^0(z)]h_\nu(z)\,dz \approx Ch_\nu(\max)B_\nu[T^0(\max)]. \qquad (5.50)$$

In general, $I_\nu^0(0) \neq I_\nu(0)$ because $T^0(z) \neq T(z)$. If we assume that C is the same in both (5.49) and (5.50), we may obtain a better estimate, $B_\nu[T^1(\max)]$, of $B_\nu[T(\max)]$ by dividing (5.49) by (5.50),

$$\frac{B_\nu[T^1(\max)]}{B_\nu[T^0(\max)]} \approx \frac{I_\nu(0)}{I_\nu^0(0)}. \qquad (5.51)$$

From $B_\nu[T_1(\max)]$ we obtain an improved estimate, $T^1(z)$, of the temperature profile.

The process may now be repeated with T^1 in place of T^0, and the process repeated until the iterations converge. This convergent solution is assumed to be the "best" solution for the temperature profile that can be obtained from the given set of radiances. Although this is a reasonable assumption, it is still an assumption, and should be tested. The results of one such test are shown in Figure 5.12. A temperature profile was assumed. Radiances were calculated from the temperature profile, and typical errors were added. From these "noisy" radiances Chahine's technique was used to recover the temperatures. As may be seen from this example, the recovery can be very good.

5.5 Reading

Further details on radiative equilibrium, radiative-convective models, and remote sounding may be found in Goody and Yung (1990), Chapter 3. A more detailed treatment of energy-balance models may be found in Lindzen (1990), Chapter 2.

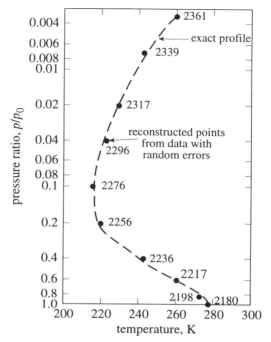

Figure 5.12 A temperature retrieval. The broken line is the original temperature profile; the points are retrieved from spectral data. The numbers are the central frequencies of the intervals used in the inversion. This exercise was theoretical, but it illustrates the process and the probable errors.

5.6 Problems

Asterisks* and double asterisks** indicate higher degrees of difficulty.

5.1 **The carbon dioxide greenhouse.** Figure 5.13 shows the outgoing radiance from Earth as measured from an orbiting satellite. The energy distribution has some resemblance to a black body at the surface temperature of the earth, T_g, but for the strong band of carbon dioxide centered on 667 cm^{-1} (for a rough approximation we may ignore the ozone band at 1042 cm^{-1}). Emission in the center of the carbon dioxide band is only about a quarter that which would occur without the gas.

Estimate the greenhouse effect caused by carbon dioxide with the following simplifying assumptions. Assume that the *only* atmospheric absorber is the 667 cm^{-1} band and that it occupies the frequency range 600 to 725 cm^{-1}. Assume that this band is represented by a rectangle and that the two sections of the spectrum on either side can be represented by triangles with bases 50 to 600 cm^{-1} and 725 to 1500 cm^{-1}. The object is to

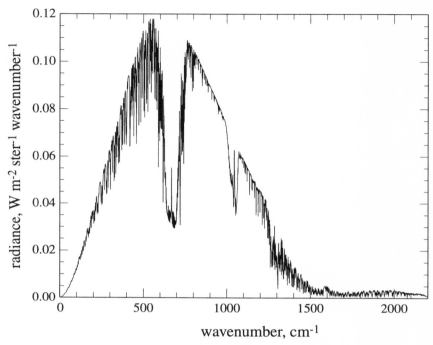

Figure 5.13 The outgoing radiance from Earth.

calculate the present ground temperature and the temperature that would occur if the carbon dioxide were removed but if all else were the same.

5.2 Radiative equilibrium in a semi-grey atmosphere.[*] A semi-grey atmosphere has two grey absorption coefficients, one for solar and one for thermal frequencies. The two radiation streams may be treated independently. In the following treatment, use the approximate equation of transfer, (4.63), neglect scattering, and assume that $\eta = \frac{\tau_S}{\tau_T}$ is a constant, where the subscripts T and S refer to thermal and solar radiation, respectively. Consider a state of radiative equilibrium.

(i) Show that the solution to (4.63) is

$$B(\tau_T) = B(0) = \frac{F_S(0)\xi_\odot}{4\pi\eta} \left[1 - \exp\left(\frac{\eta \tau_T}{\xi_\odot}\right)\right] \left(3 - \frac{\eta^2}{\xi_\odot^2}\right).$$

(ii) Show that the upper boundary condition is

$$B(0) = \frac{F_S(0)}{2\pi} \left(\frac{\eta}{2\xi_\odot} - 1\right).$$

(iii) Plot the solutions from $\tau = 0$ to 1 for $\xi_\odot = -\frac{1}{4}$ and $\eta = 0, 1, \infty$, and try to rationalize the result in your own terms.

5.3 Radiative time constants.* For any particular circumstance we may define a radiative time constant by

$$\tau_{\text{rad}}^{-1} = \frac{1}{T}\left(\frac{\partial T}{\partial t}\right)_{\text{rad}} = \frac{1}{T}\frac{\dot{q}_{\text{rad}}}{c_p}.$$

$\left(\frac{\partial T}{\partial t}\right)_{\text{rad}}$ is a virtual temperature change, and $\dot{q}_{\text{rad}}$ is the radiative heating rate per unit mass.

(i) What is the radiative time constant of an entire atmosphere of mass, M kg m^{-2}, and emission temperature, T_e?

(ii) For a planet of radius R and winds of velocity V, the order of magnitude of the dynamical response time is $\tau_{\text{dyn}} = \frac{R}{V}$. Reinterpret the argument leading to equation (5.31) in terms of competition between radiative and dynamical processes.

(iii) Calculate τ_{rad} for Earth, Mars, and Venus from the data in Table 5.2. Venus and Mars atmospheres are predominantly CO_2, for which $c_p \approx 8.5 \times 10^2$ J K^{-1}kg^{-1}.

(iv) Earth and Mars rotate at with approximately the same angular velocity, while the Venus day is 117 earth days. There are weak thermal tides in the Earth's atmosphere. Comment upon the possibility of diurnal thermal tides being excited in Mars and Venus.

5.4 Limb darkening or brightening. To a first approximation, the thermal emission from Venus has circular isophotes, that is, temperature appears to be symmetrical about the center of the field of view (see Figure 5.14). This symmetry owes its existence to the fact that temperature varies very little in any horizontal direction and may be taken to be a function of height only.

The atmosphere of Venus is so opaque in the thermal spectrum that no outgoing radiation originates at the surface. We assume that the source function varies in the vertical according to

$$B_\nu(\tau_\nu) = \alpha + \beta\tau_\nu,$$

where α and β are constants.

(i) Show that the outgoing radiance at radius, r, is

$$I_\nu^+(r) = \alpha + \beta\left(1 - \frac{r^2}{R^2}\right)^{1/2}.$$

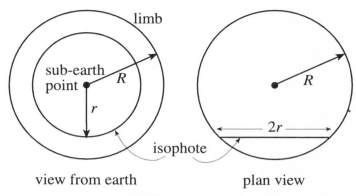

Figure 5.14 Thermal emission from Venus.

(ii) With what properties of the source function do you associate the terms *limb darkening* and *limb brightening*? Use the theory of the Chapman function to determine what optical depth your statement refers to.

(iii) On Earth, the strong Q-branch of the 667 cm^{-1} CO$_2$ band has unit optical depth close to 30 km altitude. In the wing of the band is a frequency with approximately the same brightness temperature, but unit optical depth close to 6 km altitude. If Earth were viewed from another planet, and if the temperature distribution in Figure 1.2 were the same all over the planet, would you observe limb darkening or limb brightening at these two frequencies?

5.5 **Remote sounding of temperature, I.** The 667 cm^{-1} CO$_2$ band is used for temperature soundings in the same manner as the 2348 cm^{-1} band, as discussed in §5.4. In the example discussed in this and the next problem, six channels are used, but at first we consider only channel 3, centered at 694.7 cm^{-1}.

Table 5.4 shows vertical transmissions from space down to a series of pressure levels,
$$\mathcal{T}_\nu(p) = e^{-\tau_\nu(p)},$$
in the 694.7 cm^{-1} channel. The surface pressure is 1019.8 mb; the surface temperature is 279.5 K. From the hydrostatic approximation, altitude, z, and $\ln p$ are almost proportional. Consequently, for vertical viewing ($\xi = 1$) it makes little difference to the kernel function (5.24) if we redefine it to be

$$h_\nu(\ln p, \xi = 1) = -\frac{\partial \mathcal{T}_\nu(p)}{\partial \ln p}.$$

dz must then be replaced by $d \ln p$ in equation (5.23).

CONSTRAINTS ON THE THERMAL STRUCTURE

Table 5.4 Transmissions at 694.7 cm^{-1}

Level	Pressure mb	Temperature, K	Transmission	Level	Pressure mb	Temperature, K	Transmission
	0		1.0000		150.2		0.2495
1		270.7		13		222.6	
	1.4		0.9837		188.4		0.1634
2		256.4		14		225.2	
	3.1		0.9704		233.1		0.0968
3		241.9		15		229.7	
	5.9		0.9497		284.8		0.0508
4		235.0		16		231.8	
	10.3		0.9188		344.3		0.0237
5		228.8		17		234.2	
	16.5		0.8740		412.2		0.0090
6		222.5		18		236.9	
	24.9		0.8168		489.2		0.0026
7		219.5		19		245.4	
	36.1		0.7458		575.8		0.0005
8		218.5		20		252.8	
	50.5		0.6609		673.0		0.0000
9		217.4		21		260.5	
	68.6		0.5638		781.3		0.0000
10		217.3		22		267.5	
	90.9		0.4584		901.5		0.0000
11		218.8		23		277.0	
	117.9		0.3508		1019.8		0.0000
12		220.7				279.5	

(i) At what pressure level is $h_\nu(\ln p)$ a maximum? It is only necessary to evaluate h_ν at about five levels to answer this question.

(ii) What is the upwelling radiance at the top of the atmosphere? From equation (5.23) the radiance may be written

$$I_\nu^+(0, \xi = 1) = \int_0^1 B_\nu(p)\, d\mathcal{T}_\nu(p).$$

This integral may be evaluated with sufficient precision using the trapezoidal rule and a hand calculator. However, if you program the integral it will help with the next question.

(iii) What is the brightness temperature for this radiance? What would

Table 5.5 Transmissions for channels 1, 2, 4, 5, and 6.

Pressure mb	Transmission				
	Channel 1	Channel 2	Channel 4	Channel 5	Channel 6
1.4	.8846	.9733	.9891	.9906	.9953
3.1	.7979	.9508	.9817	.9848	.9925
5.9	.7061	.9139	.9732	.9763	.9885
10.3	.6094	.8591	.9597	.9645	.9828
16.5	.5001	.7831	.9403	.9485	.9747
24.9	.3840	.6853	.9167	.9290	.9652
36.1	.2716	.5691	.8887	.9065	.9566
50.5	.1738	.4424	.8555	.8821	.9431
68.6	.0980	.3160	.8162	.8567	.9307
90.9	.0468	.2008	.7699	.8304	.9173
117.9	.0179	.1080	.7152	.8029	.9026
150.2	.0052	.0456	.6520	.7731	.8861
188.4	.0010	.0139	.5812	.7397	.8674
233.1	.0001	.0028	.5033	.7011	.8454
284.8	.0000	.0004	.4195	.6561	.8187
344.3	.0000	.0000	.3365	.6064	.7883
412.2	.0000	.0000	.2514	.5475	.7493
489.2	.0000	.0000	.1706	.4785	.6992
575.8	.0000	.0000	.1017	.3993	.6326
673.0	.0000	.0000	.0516	.3127	.5467
781.3	.0000	.0000	.0221	.2261	.4476
901.5	.0000	.0000	.0076	.1456	.3371
1019.8	.0000	.0000	.0019	.0770	.2099

it be if Chahine's first approximation, $T^0(z)$ in §5.4, were correct? You may use whichever frequency units you wish, but if you use wavenumbers (as is usual), note that this is not an MKS unit.

5.6 Remote sounding of temperature, II*. You can only do this question if you have programmed the radiance integral.

Table 5.5 gives transmission data for five additional channels in the 667 cm^{-1} CO_2 band: channel 1669.0 cm^{-1}; channel 2676.7 cm^{-1}; channel 4708.7 cm^{-1}; channel 5723.6 cm^{-1}; channel 6746.7 cm^{-1}.

Explore Chahine's inversion technique as follows. First calculate the outgoing radiance for all six channels. Channels 5 and 6 have a substantial contribution from the surface. The final temperature in Table 5.4 is the

CONSTRAINTS ON THE THERMAL STRUCTURE

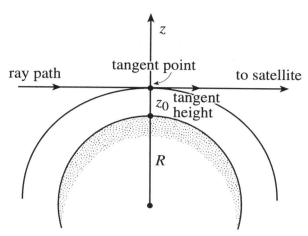

Figure 5.15 Geometry of limb viewing.

surface temperature. Review the derivation of equation (4.26). Then use Chahine's technique to derive a temperature profile, and compare it with the original. You may neglect the effect of temperature upon transmission. You will have to interpolate temperatures. Do so in terms of the layer number as an independent variable. Layers are defined by the upper and lower pressures in Tables 5.4 and 5.5.

5.7 **Limb soundings.*** Figure 5.15 shows the geometry for a satellite radiometer viewing the limb of a planet. The closest approach of the viewing ray is z_0, the *tangent height*. The optical path, $\tilde{\tau}_\nu$, is measured along the ray path. For the geometry in Figure 5.15, $\tilde{\tau}_\nu(z)$ is two-valued; we take it to be defined on the entry path only.

(i) With this definition, adapt equation (4.26) to show that the radiance recorded by the satellite is

$$I_\nu = \int_0^{\tilde{\tau}_\nu(z_0)} 2B_\nu(z) e^{-\tilde{\tau}_\nu(z_0)} \cosh\{\tilde{\tau}_\nu(z_0) - \tilde{\tau}(z)\} \, d\tilde{\tau}_\nu(z).$$

(ii) For this geometry the zenith angle varies along the path of the ray. Show that, for $z/R \ll 1$,

$$\xi(z) = \sqrt{\frac{2(z - z_0)}{R}}.$$

(iii) Assuming that the optical depth follows the exponential relation, equation (5.14), show that

$$\tilde{\tau}_\nu(z) = \tau_\nu(z_0)\sqrt{\frac{2R}{H}}\frac{\sqrt{\pi}}{2}\left\{1 - \mathrm{Erf}\left(\frac{z-z_0}{H}\right)^{\frac{1}{2}}\right\},$$

$$= \tilde{\tau}_\nu(z_0)\left\{1 - \mathrm{Erf}\left(\frac{z-z_0}{H}\right)^{\frac{1}{2}}\right\}.$$

Erf(x) is the error integral defined by

$$\mathrm{Erf}(x) = \frac{2}{\sqrt{\pi}}\int_0^x e^{-t^2}\,dt.$$

(iv) Show that the kernel function, as defined by equation (5.23), is given by

$$h_\nu(z) = \frac{\tau_\nu(z_0)}{H}e^{-\tau_\nu(z_0)}\sqrt{\frac{2R}{z-z_0}}\cosh\left\{\tilde{\tau}_\nu(z_0)\mathrm{Erf}\left(\frac{z-z_0}{H}\right)^{\frac{1}{2}}\right\}e^{-(z-z_0)/H}.$$

Calculate numerical values of $h_\nu(z)/h_\nu(z_0 + H)$ for $\tau_\nu(z_0) = 1$ and compare your results with the Chapman function in Figure 5.6. Some values for the error integral are given in the table in the solution.

CHAPTER 6

OZONE

Atmospheric ozone is a striking example of a disequilibrium chemical species, whose presence in the atmosphere was not suspected until demonstrated from measurements. The first step in the process of ozone formation is the photolysis of oxygen molecules into oxygen atoms by photons with wavelengths less than 240 nm. Because absorption at these wavelengths is strong, actinic photons do not penetrate below about 40 km. Ozone is, therefore, principally a topic of stratospheric chemistry, although the smaller amounts in the troposphere also have important environmental implications (see Chapter 7).

In the year 1930 Sydney Chapman gave an account of the stratospheric ozone layer in terms of reactions between the three oxygen species: molecular oxygen, atomic oxygen, and ozone. Atomic oxygen and ozone cycle rapidly, without any overall change in the concentration of species with odd numbers of oxygen atoms. Loss of both ozone and atomic oxygen occur when they react together to reform molecular oxygen. Chapman's theory predicts a maximum ozone concentration in the stratosphere, but the predicted concentration is larger than is observed. Since 1950 the search has been on for additional ozone loss reactions that can account for this discrepancy.

Loss reactions often require only minute concentrations of active species: They are usually catalytic, that is, the active species is not destroyed in the reaction. These active species fall into three categories: odd hydrogen compounds, odd chlorine compounds, and odd nitrogen compounds. The word odd denotes rupture of the strong bonds that bind hydrogen, chlorine, and nitrogen molecules; ultimately this rupture can be attributed to photolysis, the same agent that enables ozone to form in the first place. These reactive species can result from natural processes, but there is special interest in active nitrogen from airplane exhausts and chlorine formed from refrigerants and spray propellants. When all known gas-phase reactions are taken into account, the ozone between 20 and 50 km, except in polar regions, appears *to be close to photochemical equilibrium.*

This brings us to two more difficult aspects of the ozone problem: transport by fluid motions and nonhomogeneous catalysis by aerosol particles.

Motions are responsible for the maximum concentrations of ozone that appear in polar regions in the spring; intuition might suggest a maximum in tropical regions where insolation is greatest. Poleward pumping of ozone takes place at all times, but in the late winter there are no destructive mechanisms, and ozone concentrations increase until the sun reappears. Motions also account for most of the ozone that exist in the troposphere, where photochemical theory predicts none, and for the increased amounts of ozone that are observed in meteorological depressions.

The Antarctic ozone hole has triggered an interest in nonhomogeneous catalysis. In the extreme cold of the polar, winter stratosphere, small crystals of ice and nitric acid can form. These particles catalyze the release of chlorine from species in which it is bound in an inactive form, and they can also remove nitrogen that can react with and remove active chlorine. The net result is a rapid increase in the concentration of free chlorine when the sun returns in the spring and a precipitous decline in the ozone amount. The chlorine species under discussion are principally anthropogenic: The ozone hole is man made.

6.1 A brief history

Atmospheric ozone is exceedingly unstable; indeed, pure ozone, under laboratory conditions, is dangerously explosive. Since its presence in the atmosphere was first established, it has been a focus of attention for atmospheric chemists and a paradigm for the disequilibrium caused by photolysis.

The first suggestion that ozone is present in the atmosphere came from Hartley in 1880. Hartley sought to explain the cutoff in the solar spectrum at 310 nm, for which ozone absorption was a possibility. The matter was not settled until 1917, when Fowler and Strutt showed that certain weak bands in the spectrum of Sirius were the Huggins bands of ozone. Strutt followed up on this discovery by demonstrating in 1918, from long-path absorption spectra, that the location of the absorption was *not* in the lower atmosphere; and Cabannes and Dufay demonstrated in 1925, from an analysis of spectra of the setting sun, that most of the gas lay tens of kilometers above the earth's surface. These discoveries set the stage for the worldwide observational network of Dobson and his collaborators, and the seminal theoretical work of Chapman. The mean vertical distribution of ozone in middle latitudes is shown in Figure 6.1. The seasonal and geographical distribution of the vertically integrated column ozone amount, as measured with Dobson spectraphotometers, is shown in Figure 6.2.

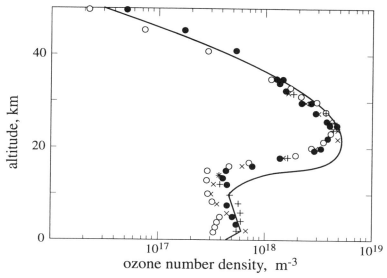

Figure 6.1 The vertical distribution of ozone in middle latitudes. The line is a theoretical calculation, and the points are measurements by a number of different observers.

Perhaps the most interesting development in the story of ozone was the work of Sidney Chapman in 1930. Chapman pointed out that the process maintaining this unstable species must be the photolysis of molecular oxygen by solar photons in the ultraviolet spectrum, and he constructed a quantitative theory based on reactions between the three oxygen species, molecular oxygen, atomic oxygen, and ozone. The theory accounted for the altitude of the peak concentration of ozone in the lower stratosphere and gave the correct order of magnitude for ozone amounts. We shall review *Chapman's theory* because it is the first of modern studies of atmospheric photochemistry and because it is still correct in some essential respects.

After World War II, academic interest in atmospheric ozone grew, and it was discovered that the recombination reactions of ozone were more complex than Chapman had suggested. In particular, Bates and Nicolet showed that catalytic recombinations involving hydroxyl radicals, derived from water vapor, played a part.

But the nature and pace of ozone research changed dramatically in the early 1970s when two independent theoretical analyses suggested that ozone was sensitive to minute concentrations of chemical species associated with human activities. These species are reactive free radicals, created when the strong molecular bonds between chlorine atoms in the chlorine molecule or between nitrogen atoms in the nitrogen molecule are ruptured by ultraviolet photons. The nitrogen bond is particularly difficult to break.

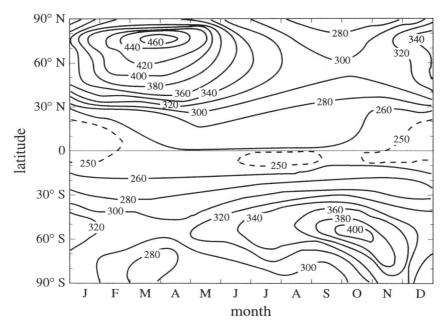

Figure 6.2 Latitude-season cross section of column ozone amounts. The column ozone amount is measured in Dobson units (DU). One DU is equal to 2.68×10^{20} molecules m^{-2}.

It can happen in internal combustion engines, and Johnston showed that the exhausts from a fleet of supersonic transports (SSTs) could strongly impact the ozone layer. The importance of chlorine was established by Rowland and Molina from their study of the fate of chlorofluorocarbons from refrigerants and spray-can propellants. Somewhat earlier the case was made that a decrease in the amount of atmospheric ozone could present a public health hazard. The subsequent environmental debate is a matter of history.

What emerges from all of these studies is the important role of free radicals having ruptured bonds of oxygen, nitrogen, hydrogen, or chlorine. As long as these ruptured bonds exist, regardless of their state of combination, many reactions, including ozone catalysis, can take place. Consequently, we consider families of *odd oxygen, odd hydrogen, odd nitrogen,* and *odd halogen compounds*, each containing many species, that interact among themselves and between families until nonreactive compounds are produced. In addition, some stable species take part in the reactions, for example, O_2, H_2O_2, HNO_3, CH_4, HCl, $ClONO_2$, H_2O, N_2O, N_2O_5, and H_2. A "modern" ozone calculation may involve 50 or more chemical species, 500 or more gas phase reactions, and a dynamical model for transport of species.

6.2 Photodissociation rates

At the heart of our discussion is the ability of energetic solar photons with wavelengths less than 242 nm to rupture the strong O–O bond in molecular oxygen. In terms of the discussion following Table 1.4, the sequence of events is

$$\begin{array}{rcl} O_2 + \gamma & \to & O_2^* \\ O_2^* & \to & O + O \\ \hline \text{net} \quad O_2 + \gamma & \to & O + O \,. \end{array} \quad (6.1)$$

γ is a solar photon. The decay of the excited state is rapid, and the reaction that will be observed is the *net* bimolecular reaction. From equation (H.2),

$$\frac{d[\gamma]}{dt} = \frac{d[O_2]}{dt} = -\frac{1}{2}\frac{d[O]}{dt} = -J_{O_2}[O_2] \,. \quad (6.2)$$

The square brackets indicate concentrations. J_{O_2} is the *photochemical loss rate* of molecular oxygen per oxygen molecule or the *inverse lifetime* of an oxygen molecule for photodissociation.

The rate of change of photon density is the rate of change of radiant energy density divided by $h\nu$. Returning to the derivation of equation (4.6), we may show that

$$\begin{aligned} \frac{d[\gamma]}{dt} &= \int_{\nu_0}^{\infty} \nabla \cdot \vec{F}_\nu \frac{d\nu}{h\nu} \,, \\ &= \int_{\nu_0}^{\infty} \frac{d\nu}{h\nu} \int_{4\pi} [O_2] e_\nu (J_\nu - I_\nu) \, d\omega_l \,. \end{aligned} \quad (6.3)$$

ν_0 is the minimum frequency that can give rise to a dissociation, corresponding to 242 nm for molecular oxygen.

The thermal source function may be neglected in the ultraviolet spectrum. From equation (4.8), the scattering source function, (4.15), and the normalization of the phase function, (3.34), (6.3) becomes

$$\frac{d[\gamma]}{dt} = -[O_2]\int_{\nu_0}^{\infty}\frac{d\nu}{h\nu}\int_{4\pi} k_\nu I_\nu \, d\omega_l. \quad (6.4)$$

The scattering coefficient does not appear explicitly in this loss process, although scattering can influence the ambient radiance, I_ν (we return to this point below).

Hence, from (6.2),

$$J_{O_2} = \int_{\nu_0}^{\infty}\frac{d\nu}{h\nu}\int_{4\pi} k_\nu I_\nu \, d\omega_l \,. \quad (6.5)$$

If, as is commonly the case, only the direct solar beam is considered, we may use equation (4.17) to give

$$J_{O_2,d}(z) = \int_{\nu_0}^{\infty} k_\nu f_\nu(z)\frac{d\nu}{h\nu} \,. \quad (6.6)$$

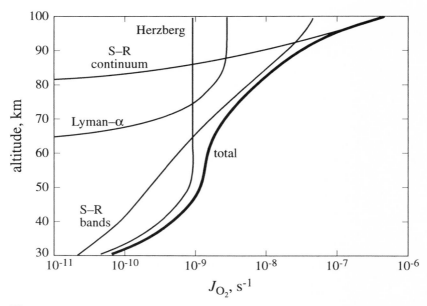

Figure 6.3 Contributions of regions to oxygen photodissociation. $\mu_\odot = 1$. The regions are identified in Figure 3.1.

$f_\nu(z)$ is the irradiance (see equation 4.19). Although oxygen has been used as an example to derive these equations, they apply, with other coefficients, to any other species.

The state of excitation of the products of photodissociation can be important, and this depends upon the frequency of the absorbed photon. For this and for other reasons, it is convenient to divide the spectrum into regions having common properties: the Herzberg continuum, the Schumann-Runge bands and continuum, and Lyman-α for oxygen (see Figure 3.1); the Chappuis, Hartley, Huggins bands, and the far ultraviolet for ozone (see Figure 3.2). For each spectral region there is a partial loss rate. These must be added to give the inverse lifetime of the molecule.

Figures 6.3 and 6.4 show partial rates for oxygen and ozone. For both gases, the most important spectral region changes with altitude, a fact that has important consequences for atmospheric chemistry. When a curve in either figure runs vertical it means, from equation (4.19), that the optical depth is small in the spectral region concerned. At low altitudes, the inverse lifetimes of oxygen and ozone are dominated by the weakest relevant absorptions: the Herzberg continuum for oxygen and the Chappuis bands for ozone.

Figures 6.3 and 6.4 were constructed from equation (6.6) for the direct

OZONE

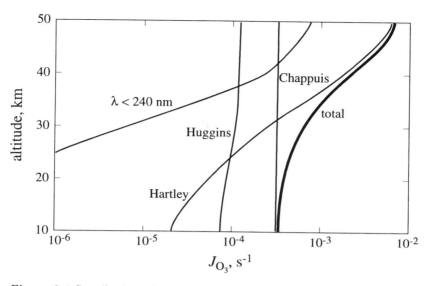

Figure 6.4 Contributions of regions to ozone photodissociation. $\mu_\odot = 1$. The regions are identified in Figure 3.2.

solar beam only. Scattering is strongest in the lower atmosphere, where most molecules, clouds, and aerosols are concentrated. Bands that are strongly absorbed before solar radiation reaches the troposphere, such as the Hartley bands of ozone, are affected by scattering only to a minor extent. The Chappuis bands of ozone, on the other hand, are affected by scattering, which adds to the radiance and increases J_{O_3}. The integral of the radiance over all angles is increased by the backscattered radiation. From equation (6.5), J_{O_3} for the Chappuis bands is, to first order, increased by a factor $\approx (1+\overline{A})$, where $\overline{A}$ is the mean planetary albedo for this spectral region.

When using the data in Figures 6.3 and 6.4, it is important to remember that the photochemical rates, as expressed by equation (6.5), are not usually simple functions of the absorber amounts at the level under consideration. In one circumstance the relationship is simple. As $\tau_\nu \to 0$, $J_{O_2} \to$ constant (see equations 6.6 and 4.19). This situation is illustrated in Figure 6.3 by the Herzberg continuum above 60 km and in Figure 6.4 by the Chappuis bands at all levels. If τ_ν is not small, however, the partial rate is a function of the absorber density at all higher levels. Moreover, in certain spectral intervals, for example, near 200 nm, where the Herzberg continuum of oxygen overlaps the Hartley band of ozone, the partial rate for each gas depends upon the densities of both gases at all levels above the level of interest.

If $D_\nu\, d\nu$ is the rate of photolysis of oxygen molecules for the spectral range ν to $\nu + d\nu$, equations (6.2), (6.6), and (4.19) lead to

$$D_\nu(z) = [O_2] k_\nu f_{\nu,0} \exp\left(-\int_z^\infty [O_2] k_\nu \frac{dz'}{\mu_\odot}\right). \tag{6.7}$$

If the optical depth for oxygen follows an exponential decrease with height, as we assumed in the discussion in §5.1 (see equation 5.14) and if the absorption coefficient is constant, equation (6.7) is, apart from a constant factor, the same as equation (5.26). The ratio $\frac{D_\nu(z)}{D_\nu(\text{max})}$ is, therefore, distributed in a Chapman layer (see Figure 5.6), with

$$D_\nu(\text{max}) = \frac{f_{0,\nu} \mu_\odot}{eH} \tag{6.8}$$

and

$$\tau_\nu(\text{max}) = [O_2](\text{max}) k_\nu H = \mu_\odot. \tag{6.9}$$

Because $[O_2]$ decreases as height increases, equation (6.9) tells us that $z(\text{max})$ increases with k_ν and decreases with $\mu_\odot$. The height of the Chapman layer is greatest for the largest absorptions, and it increases as the sun goes down.

As a historical footnote, Chapman gave the theory of his eponymous layer in the present context and not in connection with thermal emission, as we introduced it. The theory happens to be the same in the two cases.

The level of $z(\text{max})$ is an important matter in any discussion of atmospheric photochemistry because this is the level at which the relevant photodissociation is most important. Figure 6.5 shows $z(\text{max})$ for a vertical sun for all wavelengths in the ultraviolet spectrum that concern this book.

6.3 Photochemistry of an oxygen atmosphere

6.3.1 Chapman's theory

Chapman considered an atmosphere in which oxygen was the only reactive species. The important reactions below 70 km were

$$O_2 + \gamma \rightarrow O + O\ (J_2), \tag{6.10}$$

$$O_3 + \gamma \rightarrow O_2 + O\ (J_3), \tag{6.11}$$

$$O + O_2 + M \rightarrow O_3 + M\ (k_2), \tag{6.12}$$

$$O + O_3 \rightarrow O_2 + O_2\ (k_3). \tag{6.13}$$

Conventional symbols for photochemical rates and reaction rate constants are given in parentheses by each of the above equations; their values are

OZONE

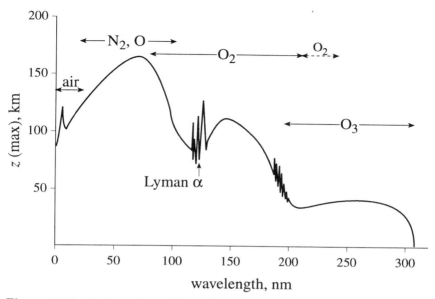

Figure 6.5 Level of maximum dissociation rate. $\mu_\odot = 1$. An average ozone profile has been used. Between 200 and 240 nm both ozone and oxygen contribute to the dissociation rate.

Table 6.1 Rate constants and photochemical rates for oxygen

Rate	Numerical value	Units
$J_2(J_{O_2})$	see Figure 6.3	s^{-1}
$J_3(J_{O_3})$	see Figure 6.4	s^{-1}
k_2	$5.6 \times 10^{-46} \left(\frac{300}{T}\right)^{2.36}$	$m^6\ s^{-1}$
k_3	$2.0 \times 10^{-17} \exp\left(-\frac{2280}{T}\right)$	$m^3\ s^{-1}$

shown in Table 6.1. M represents any molecule, the function of which is to carry away energy and momentum. Two particles cannot fuse into one and simultaneously satisfy conservation of both energy and momentum. In the absence of a third body, reaction (6.12) cannot take place.

Equations (6.10) to (6.13) can be solved for $[O_2]$, $[O_3]$, $[O]$, and $[\gamma]$ by numerical methods using the data given in Table 6.1. J_2 and J_3 depend upon concentrations of oxygen and ozone above the level of interest, but they are constant at sufficiently high levels, and the equations can be solved iteratively, starting from the "top" of the atmosphere and working

Table 6.2 Measured concentrations of oxygen species at low latitudes

Altitude, km	$[O_2]$, m^{-3}	$[O_3]$, m^{-3}	$[O^3P]$, m^{-3}	$[O^1D]$, m^{-3}
10	1.7 (24)	1.0 (18)	1.3 (10)	—
15	8.1 (23)	1.1 (18)	5.5 (10)	—
20	3.6 (23)	2.9 (18)	9.4 (11)	9.0(5)
25	1.6 (23)	3.2 (18)	6.7 (12)	5.0(6)
30	7.1 (22)	3.0 (18)	1.8 (13)	2.5(7)
35	3.5 (22)	2.0 (18)	2.4 (14)	1.0(8)
40	1.7 (22)	1.0 (18)	1.2 (15)	3.3(8)
45	8.9 (21)	3.2 (17)	3.7 (15)	6.0(8)
50	4.8 (21)	1.0 (17)	6.5 (15)	6.1(8)
55	2.6 (21)	3.2 (16)	8.4 (15)	4.4(8)
60	1.5 (21)	1.0 (15)	6.5 (15)	2.6(8)
65	8.2 (20)	3.2 (15)	5.0 (15)	1.5(8)
70	4.2 (20)	1.0 (15)	4.0 (15)	9.6(7)

Powers of 10 are given in parentheses. These data are averages of what are, except for $[O_2]$, highly variable quantities.

downwards.

To obtain more direct insight into the nature of these reactions, we must make approximations based upon the magnitudes of observed concentrations of oxygen species (see Table 6.2 for data below 70 km). First, we note that almost all oxygen below 70 km is in the form of O_2,

$$2[O_2] \gg [O] + 3[O_3] \ . \tag{6.14}$$

As a consequence, below 70 km $[O_2]$ may be treated as an independent variable.

The rate equations for atomic oxygen and ozone are

$$\frac{d[O]}{dt} = 2J_2[O_2] + J_3[O_3] - k_2[O][O_2][M] - k_3[O][O_3] \tag{6.15}$$

and

$$\frac{d[O_3]}{dt} = -J_3[O_3] + k_2[O][O_2][M] - k_3[O][O_3] \ . \tag{6.16}$$

A numerical analysis of the terms in (6.15) and (6.16), using the data in Tables 6.1 and 6.2, shows that the dominant terms on the right sides of both equations are $J_3[O_3]$ and $k_2[O][O_2][M]$. This implies that the rates of (6.11) and (6.12) are much faster than the rates of (6.10) and (6.13), and that the former two reactions take place many times before the latter two take place once. But, as regards oxygen species, (6.11) and (6.12) are the reverse of each other: Equation (6.11) destroys a molecule of ozone and creates one atom of oxygen; while (6.12) destroys an atom of oxygen and creates one molecule of ozone. A strong feature of these equations is a rapid *internal cycling* between O and O_3. There is a balance between the two dominant terms, represented by

$$\frac{[O]}{[O_3]} = \frac{J_3}{k_2[M][O_2]} . \qquad (6.17)$$

If this balance is perturbed it is, for the altitude range under discussion, rapidly restored. The only relevant question is the fate of total odd oxygen, $[O_x]=[O]+[O_3]$. The common property of odd oxygen species, O_x, is that an O–O bond has been ruptured.

Adding equations (6.15) and (6.16) gives the rate equation for odd oxygen:

$$\frac{d[O_x]}{dt} = 2J_2[O_2] - 2k_3[O][O_3] . \qquad (6.18)$$

Further numerical analysis shows that, below 75 km

$$\frac{k_3[O_3]}{k_2[O_2][M]} < 10^{-3},$$

so that (6.16) may be simplified to

$$\frac{d[O_3]}{dt} = -J_3[O_3] + k_2[O][O_2][M] . \qquad (6.19)$$

Equations (6.17), (6.18), and (6.19) are the fundamental equations of Chapman's theory for altitudes less than 75 km.

For a steady state, $\frac{d[O_x]}{dt} = \frac{d[O_3]}{dt} = 0$. The photochemical equilibrium solution for ozone is

$$[O_3]_e^2 = \frac{J_2 k_2}{J_3 k_3}[O_2]^2[M] . \qquad (6.20)$$

$[O_3]_e$, as given by equation (6.20), has a maximum in the stratosphere. At low levels, $[O_3]$ is small because J_2 is small (see Figure 6.3); at high levels, on the other hand, $[O_3]$ is small because $[O_2]$ and $[M]$ are both small.

Calculations based on equation (6.20) are shown in Figure 6.6, where they are compared with some relevant observations. The theory gives a maximum concentration near to 25 km, where it is observed, and its magnitude is correct to within a factor of three. These are remarkable achievements for such a simple theory, but the discrepancies are large enough to show that the physics is incomplete. Specifically, throughout the altitude

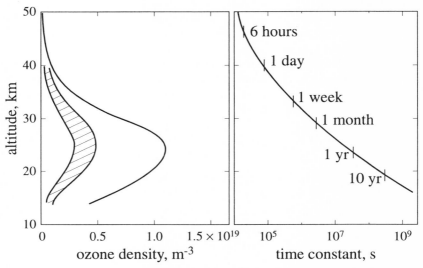

Figure 6.6 Calculated and observed ozone concentrations and time constants. The isolated line in the left-hand panel is calculated from the equilibrium equation (6.20). The shaded area gives the range of observed data at low latitudes. The right-hand panel will be discussed in the next section. The time constant plotted is one half that given by equation (6.21).

range represented in Figure 6.6, the predicted concentration is too high, while in the troposphere (not shown in the figure) it is too low. A substantial concentration of ozone is observed in the troposphere (see Figure 6.1), but Chapman theory requires negligible values because J_2 is very small at low levels.

After nearly 70 years of attempts to improve Chapman's theory, it has survived remarkably well. The ozone source, equation (6.12), is essentially correct. On the other hand, the ozone sink, equation (6.13), is too slow because Chapman did not anticipate catalysis by odd hydrogen, odd nitrogen, and odd halogen species (see §6.4). Finally, the theory gives no useful information on tropospheric ozone, which is mainly governed by fluid transports. We shall discuss these limitations in the following sections.

6.3.2   Excited states of atomic and molecular oxygen

The two photolytic process (6.10) and (6.11) can, for sufficiently energetic photons, yield an oxygen atom in an excited ^1D state. This state has an energy 2 eV higher than that of the ground state, ^3P, a substantial energy difference in terms of activation energies of atmospheric reactions, and it can make a very large difference to the chemistry. The excited state can decay spontaneously to the ground state with the emission of a doublet at 636.6 and 630.0 nm. This red doublet is prominent in the faint light emitted from the night sky.

Reactions (6.11) and (6.13) can both give rise to excited molecular oxygen in the $^1\Delta_g$ state, which lies about 1 eV above the ground, $^3\Sigma_g^+$, state. Radiative transitions between these states give rise to photons of wavelength 1.27 μm, which are also observed in the light from the night sky. The energy of the excited state of molecular oxygen is also sufficient to influence rates of reactions, but not as strongly as $O(^1D)$.

6.3.3 Departures from equilibrium

In §6.3.1 we noted that tropospheric ozone is influenced by fluid transports. In the spirit of the discussion of thermal equilibrium in §1.4.3, we may examine this question to order of magnitude by comparing characteristic times for photochemistry and dynamics; the observed state should be closer to the state with the shorter relaxation time.

Because odd oxygen species cycle rapidly, the relevant chemical time constant is that for total odd oxygen, τ_{O_x}. With the approximation $[O_x] \approx [O_3]$ (see Table 6.2), we may obtain this time constant from equation (6.18). The first term on the right does not depend upon odd oxygen and will not influence the time to restore equilibrium. From equation (H.7) the chemical time constant is

$$\tau_{O_x} = \frac{1}{2k_3[O]} \, . \tag{6.21}$$

We now assume that we are close to photochemical equilibrium so that, from equations (6.17) and (6.20),

$$\tau_{O_x} \approx \frac{1}{2} \left(\frac{k_2[M]}{k_3 J_2 J_3} \right)^{0.5} . \tag{6.22}$$

Equation (6.22) is plotted in Figure 6.7. The range of chemical time constants is large, from years or more below 20 km, to hours or minutes between 40 and 80 km.

Transport can be either by large-scale motions or by small-scale eddies, and both can compete with the chemistry. As far as eddies are concerned, the characteristic transport time will be discussed in Chapter 9, with respect to equation (9.37). Vertical eddy diffusion coefficients are given in Appendix C, Figure C.4. Calculations based on these coefficients are indicated by τ(eddy) in Figure 6.7.

Large-scale transport times may be estimated from a characteristic distance scale divided by a velocity. The magnitudes of global wind velocities are known from observation; see Appendix C for a discussion of climatological mean winds: zonal (u), meridional (v), and vertical (w). To order

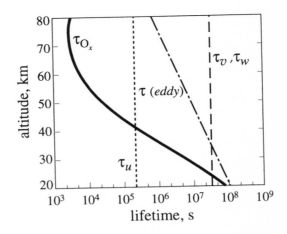

Figure 6.7 Orders of magnitude of photochemical and transport for ozone. τ_u is based on $u = 30$ ms^{-1}, τ_v on $v = 0.2$ ms^{-1} and, to order of magnitude, $\tau_w = \tau_v$.

of magnitude, the horizontal scale for global motions is the radius of the planet, R_e, and the vertical scale is the scale height, H. Hence,

$$\tau_u \sim \frac{R_e}{u}, \text{ for zonal winds },$$
$$\tau_v \sim \frac{R_e}{v}, \text{ for meridional winds }, \quad (6.23)$$
$$\tau_w \sim \frac{H}{w}, \text{ for vertical winds }.$$

These time constants are also shown in Figure 6.7.

Despite the many approximations involved, the data in Figure 6.7 give important information on the mechanisms that lie behind the observed ozone amounts. Above 40 km, the shortest lifetime is the photochemical lifetime; in this region of the atmosphere we may expect, on average, an approach to photochemical equilibrium, and this is observed, more or less, to be the case. Between 20 and 40 km, zonal motions provide the only competition with photochemistry. Zonal winds blow from day to night. When they are important they have the effect of smoothing out day-night differences but will not affect meridional gradients. In this region of the atmosphere we anticipate that the ozone will behave as if in equilibrium, but with the diurnal average insolation. Below 20 km every class of motion will perturb the photochemistry, and photochemical equilibrium is irrelevant. We may anticipate instead a tendency toward a mixed atmosphere in the troposphere, with somewhat similar ozone concentrations at all levels. Again, this is observed to be the case.

6.4 Gas phase catalysis

The two processes involving the hydroxyl molecule (OH) that were discussed by Bates and Nicolet were

$$\begin{array}{rrcl} & OH + O_3 & \to & HO_2 + O_2 \\ & HO_2 + O & \to & OH + O_2 \\ \hline net & O_3 + O & \to & 2O_2 \, , \end{array} \quad (6.24)$$

effective in the upper stratosphere (30 to 50 km) and in the lower stratosphere (18 to 25 km), where there is little atomic oxygen,

$$\begin{array}{rrcl} & HO_2 + O_3 & \to & OH + 2O_2 \\ & OH + O_3 & \to & HO_2 + O_2 \\ \hline net & 2O_3 & \to & 3O_2 \, . \end{array} \quad (6.25)$$

OH and HO_2 both belong to the odd hydrogen family. They are rapidly cycled in the above reactions and are conserved, so that their role is catalytic.

The hydrogen bond in water or methane may be broken by photolysis,

$$H_2O + \gamma \to H + OH, \; \lambda < 200 \text{ nm} \quad (6.26)$$

or by reactions with excited oxygen atoms,

$$H_2O + O(^1D) \to 2OH \quad (6.27)$$

and

$$CH_4 + O(^1D) \to CH_3 + OH \, . \quad (6.28)$$

Reactions (6.27) and (6.28) do not take place if the atomic oxygen is in the ground state, but the added excitation energy of the ^1D state is sufficient to overcome the activation energy of the reaction. The photolytic reaction (6.26) requires energetic photons with wavelengths less than 200 nm. Photons of this wavelength do not penetrate below 40 km (see Figure 6.5). At lower levels, the reactions (6.27) or (6.28) with atomic oxygen are more important than photolysis.

All of the known catalytic radicals undergo complex reaction chains. The odd hydrogen reactions are shown schematically in Figure 6.8. This figure is not explicit and requires interpretation by the reader. For example, the far left-hand balloons and arrows represent (6.27) and (6.28), while the arrow from HO_2 to H_2O_2, involving HO_2, must be construed as

$$HO_2 + HO_2 \to H_2O_2 + O_2 \, , \quad (6.29)$$

and so on. Note that Figure 6.8 involves some reactions with odd nitrogen and odd chlorine species; the chemistries of the catalytic families are not independent.

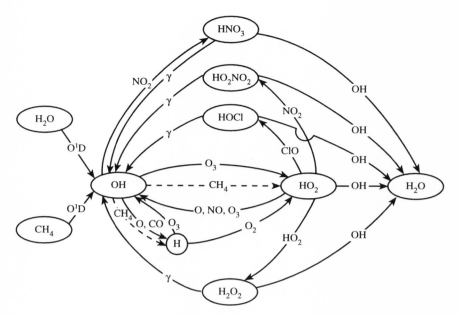

Figure 6.8 Reactions involving odd hydrogen species.

An important issue in these reaction chains is the removal of catalytic species, or chain termination (compare §1.5.2). Hydrogen peroxide (H_2O_2) is relatively stable, so that reaction (6.29) has the effect of removing from immediate consideration two members of the reactive, odd hydrogen family. A chain termination reaction for *both* odd hydrogen and odd nitrogen is

$$OH + NO_2 + M \rightarrow HNO_3 + M \,. \tag{6.30}$$

Nitric acid (HNO_3) is stable, soluble in water, and can be rained out of the atmosphere.

The odd nitrogen and odd chlorine families are, respectively (N, NO, NO_2, $2N_2O_5$, HNO_3) and (Cl, ClO, HCl, HOCl). Their catalytic reactions with odd oxygen are analogous to (6.24) and (6.25); for example,

$$\begin{array}{rrcl} & NO + O_3 & \rightarrow & NO_2 + O_2 \\ & NO_2 + O & \rightarrow & NO + O_2 \\ \hline net & O_3 + O & \rightarrow & O_2 + O_2 \end{array} \tag{6.31}$$

and,

$$\begin{array}{rrcl} & Cl + O_3 & \rightarrow & ClO + O_2 \\ & ClO + O & \rightarrow & Cl + O_2 \\ \hline net & O_3 + O & \rightarrow & O_2 + O_2 \,. \end{array} \tag{6.32}$$

Reactions (6.31) first came to the attention of atmospheric scientists in connection with the effects of the exhaust emissions from a fleet of supersonic transports. That is not a current environmental concern, but the reactions may also be important because of nitrous oxide (N_2O) emitted from agriculture and mixed up into the stratosphere. Nitrous oxide can react with excited atomic oxygen to give odd nitrogen,

$$N_2O + O(^1D) \rightarrow NO + NO \ . \qquad (6.33)$$

Nitrous oxide is a stable, well-mixed constituent of the atmosphere (see Table 1.1). It is formed by microbial reactions in the soil and in the surface layers of the ocean; it is destroyed photochemically in the stratosphere. Its lifetime in the atmosphere is about 150 years. Fertilization of the land by man and by animals enhances the production of nitrous oxide, which is currently increasing at a rate of about 0.25% per year.

Of more immediate concern are the chlorine atoms produced in the stratosphere by photolysis of both natural and anthropogenic chlorine compounds. Hydrogenated chlorine compounds can also produce odd chlorine in reactions with OH and $O(^1D)$. The source of concern is the emission of industrial chlorofluorocarbons (CFCs, for short, are halogenated methanes with the general formula $C_nF_xCl_{4n-x}$). CFCs are unreactive in the troposphere, and their concentration could build up indefinitely were it not for slow mixing into the stratosphere, where they can be broken down by photolysis; for example, CFC–12 gives

$$CF_2Cl_2 + \gamma \rightarrow CF_2Cl + Cl, \ \lambda < 215 \text{ nm} \ . \qquad (6.34)$$

The residence time of CFC–12 in the atmosphere is about 150 years. Once a reservoir of CFCs has built up in the troposphere large enough to supply chlorine in dangerous concentrations to the stratosphere, the problem will remain for at least the next century, even if nothing further is added.

It has been pointed out that the catalytic families can interact. A particularly interesting reaction (because it involves both odd chlorine and odd nitrogen) terminates with the relatively stable species chlorine nitrate,

$$ClO + NO_2 + M \rightarrow ClONO_2 + M \ . \qquad (6.35)$$

In the gas phase, this reaction can only be reversed by a very slow photolysis; but see §6.7 for a surprising aspect of this question.

The overall position with respect to the catalytic destruction of ozone between 20 and 50 km is illustrated in Figure 6.9. These rates were calculated by applying the principal reactions to the *observed* concentrations of odd oxygen, odd hydrogen, odd nitrogen, and odd chlorine. Photolysis of molecular oxygen, equation (6.10), is the source of odd oxygen (broken line

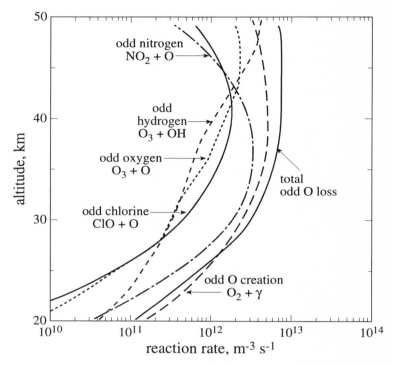

Figure 6.9 Odd oxygen creation and destruction in middle latitudes. The two furthest right curves show the rate of production of odd oxygen (broken line) and its total loss rate (full line). In the four left-hand curves, the loss rate is broken down into reactions between odd oxygen and odd oxygen, odd hydrogen, odd nitrogen, or odd chlorine.

in Figure 6.9). Equation (6.13) is the loss mechanism for odd oxygen in the Chapman scheme; other oxygen reactions preserve O_x. Equations (6.24) and (6.25) are used to calculate the loss from odd hydrogen reactions, while (6.31) and (6.32) give the odd nitrogen and the odd chlorine reactions.

The reader is reminded that the analysis in Figure 6.9 is not based on a predictive theory but depends on the availability of observed data on species such as NO_2 and ClO. Total odd oxygen loss and production are close to balance: This region of the atmosphere is at most latitudes close to photochemical equilibrium. The Chapman reactions by themselves would require a balance between total production and the "O_3+O" curve, confirming the commentary on Figure 6.6 to the effect that this loss term is insufficient by itself.

6.5 Transport

Chemical equations, such as (6.15) and (6.16), contain no term for the expansion or contraction of the system; in effect, they apply to a system with constant overall density. In the absence of chemical reactions, however, it is the mixing ratio ($\propto \frac{[X]}{\rho}$) rather than the number density, that is conserved along the flow. We may meet this objection by writing these equations in the form

$$\frac{d\left(\frac{[X]}{\rho}\right)}{dt} = \frac{P_X}{\rho} - \frac{D_X}{\rho}, \tag{6.36}$$

where P_X represents the *production* or positive terms on the right side of (6.15) or (6.16), and D_X represents the *destruction* or negative terms.

Expanding the complete differential and making use of the equation of continuity, (F.1), we obtain the equation of continuity for the species, X,

$$\underbrace{\frac{\partial [X]}{\partial t}}_{rate\ of\ change} + \underbrace{\nabla \cdot \vec{F}_X}_{transport} = \underbrace{P_X}_{production} - \underbrace{D_X}_{destruction}, \tag{6.37}$$

where

$$\nabla \cdot \vec{F}_X = \vec{u} \cdot \nabla [X] + [X] \nabla \cdot \vec{u}. \tag{6.38}$$

[X] can change locally if the velocity diverges, so that the density changes, or if there is flow along a gradient of the concentration.

Fluxes can conveniently be divided into fluxes on the mean flow and fluxes due to small-scale motions (see §9.1.3 for more discussion). When calculations are performed with a modern general circulation model (GCM), the division between these fluxes lies at the grid scale of the calculation. Motions larger than this scale are calculated explicitly, even though they may have complex features; motions on the subgrid scale are treated as diffusive. We previewed the time constants for mean motions, eddy motions and photochemistry in Figure 6.7. The conclusions from these order-of-magnitude arguments are valid, but for more precise discussions it is usual to use a GCM with the equation of continuity, (6.37).

The classic case of transports affecting a chemical species is the observed relationship between ozone and synoptic-scale weather systems. The ozone amount is observed to increase in low-pressure systems and to decrease in high-pressure systems. A well-documented example is shown in Figures 6.10(a) and (b).

The 300 mb geopotential heights[1] in Figure 6.10(b) show the storm

---
[1] Figure 6.10(b) displays the relative height of a constant-pressure surface. The height is calculated from the hydrostatic equation using a standard value for gravity and is called *geopotential height*. The more familiar way to represent pressure data is as pressure contours at constant height. Meteorologists tend to prefer the former method. Apart from scale, the two representations are interchangeable.

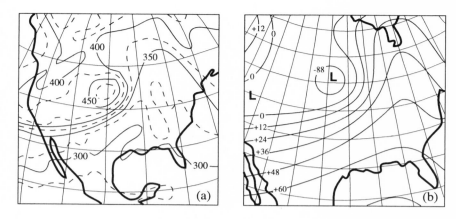

Figure 6.10 Ozone in a depression. (a) Isopleths of total ozone amount in Dobson units recorded on April 28, 1984. (b) 300 mb geopotential heights at approximately the same time. L indicates low pressure.

position, while Figure 6.10(a) shows an increase in the column amount of ozone of nearly 50% at the center of the storm. The vertical motion in the core of a depression is downward at high altitudes, while the concentration of ozone increases with height. The first term on the right side of equation (6.38) is then negative, and $[O_3]$ will increase with time. Alternatively, we may see this as a process transferring ozone from high levels, where it can be replaced in days (see Figure 6.7) to the troposphere, where its chemical lifetime is longer than 10 years.

Another important example of the effect of transports on ozone concentration is to be found in the seasonal maximum of ozone that lies close to 75° in the northern spring (see Figure 6.2). When Chapman's theory was first advanced, the fact that the maximum did not correspond to the maximum insolation (in the tropics) was cited as an objection to a photochemical theory. Increased insolation does not necessarily lead to higher ozone amounts. The insolation affects J_3 (ozone destruction) as well as J_2 (ozone production), and it is the ratio of these two quantities that determines the equilibrium ozone amount (equation 6.20). Moreover, high latitudes are generally colder than low latitudes, and the effect of temperature on k_3 (see Table 6.1) is such that a decrease of temperature from 265 to 235 K increases the photochemical equilibrium ozone amount up to 40%.

Under these circumstances it is difficult to construct a simple deductive argument that adequately describes the latitudinal and seasonal changes of ozone. Instead, we shall look at results from a general circulation model that employed a fairly complete set of chemical equations to calculate the ozone, including odd hydrogen, odd nitrogen, and odd chlorine catalysis.

OZONE 179

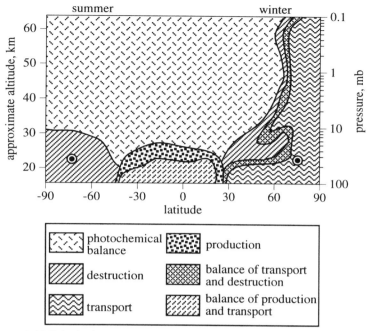

Figure 6.11 Principal terms in the ozone balance of the stratosphere. See the text for explanation of the terms. The circled points will be explained in connection with Figure 6.12.

Figure 6.11 shows schematically the two dominant terms in the continuity equation for ozone (see equation 6.37). In the three cases for which only one term is mentioned, the balance is between that and the rate of change term, $\frac{\partial [O_3]}{\partial t}$. In the other three cases, where the term *balance* is employed, the named terms dominate and there is approximate balance between them; in the case of *photochemical balance*, the two terms are the chemical production and destruction terms, P_{O_3} and D_{O_3}.

Two immediate conclusions may be reached from inspection of Figure 6.11. First, except in the polar winter, the atmosphere between 25 and 65 km is in photochemical equilibrium. This conclusion agrees with that reached in connection with Figure 6.7. Second, poleward of 60° in the winter hemisphere, chemical production and destruction terms are small; the dominant terms are rate of change and transport, and this is where we must look for an explanation for the spring maximum in northern latitudes.

Some details at 43 mb and 75° N are shown in Figure 6.12. The first panel compares the observed and calculated ozone mixing ratios and establishes that the model does reproduce the observed spring maximum

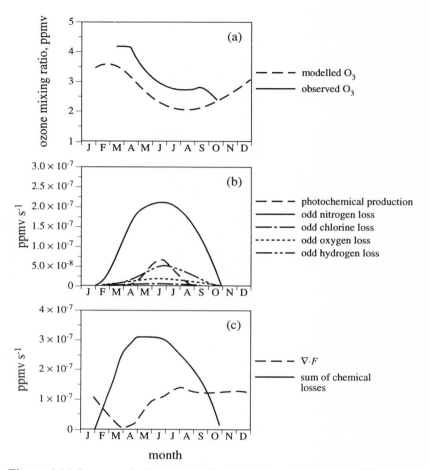

Figure 6.12 Ozone production, destruction, and transport terms at 75°N and 42 mb. (a) Observed and calculated ozone mixing ratios (see Appendix A for definition of mixing ratios). (b) Chemical production and destruction rates for odd oxygen. (c) Transport and chemical terms. The rate of change is not given explicitly, but it is the difference between the terms in (c). This figure uses the same data as were used in Figure 6.11. The location of these calculations is indicated by the circled points in Figure 6.11.

approximately. The second panel shows the chemical terms, establishes that there is no important chemical activity during the winter and shows that, during the summer, the dominant chemical process is destruction by odd nitrogen catalysis. The third panel shows the chemical terms combined and the transport term from equation (6.37); the difference between these two equals the rate of change, $\frac{\partial [O_3]}{\partial t}$.

With this information we can put together an explanation of the spring ozone maximum. Starting in November, the dominant chemical destruction term, destruction by odd nitrogen catalysis, starts to fall and remains low until the next February. The reason for this is that, in the absence of sunlight, there is no mechanism for forming the free radicals that serve as the fuel for reactions forming odd nitrogen, such as equation (6.33). During the winter, transport of ozone from low latitudes continues, and because it is not destroyed, its concentration grows steadily. The process stops when sunlight starts to form odd nitrogen in February, and rapid catalytic destruction begins. The net result is a maximum ozone concentration during February and March.

6.6 Tropospheric ozone

Vertical diffusion coefficients (see Figure C.4) are much lower at and just above the tropopause than they are in the troposphere. For this reason, it is common practice to treat the troposphere as a box, having the tropopause as a lid, with internal chemistry and reactions at the earth's surface. In this section we consider only tropospheric ozone, although ozone does not act independently from other tropospheric oxidizers, such as $O(^1D)$ and OH, see §7.1.

The photolytic source of odd oxygen falls to very small values in the troposphere (see Figure 6.3). Nevertheless, ozone concentrations in the troposphere are not small (see Figure 6.1), about 5×10^{17} m^{-3}, for a worldwide total of about 260 Tg^2 of trophospheric ozone.

Ozone is destroyed at the earth's surface. The destruction rate has been uncertainly estimated to be about 1000 Tg y^{-1}, giving a lifetime with respect to this process alone of about 100 d. To maintain a steady state we need to identify a source of about this magnitude. One such source, suggested by the order-of-magnitude arguments of §6.3, is downward transport from the stratosphere. Here again, estimates are very uncertain, but the net transport to the troposphere by motions on every scale has been estimated to be about 600 Tg y^{-1}.

Chemical formation and destruction terms in the troposphere are also hard to estimate. Ozone can be destroyed photolytically in the troposphere by process (6.11), but the atomic oxygen formed usually reacts with molecular oxygen to form ozone again (equation 6.12); this is one aspect of the cycling of O_x. However, most of the atomic oxygen formed in ozone photolysis is in the excited ^1D state; this species reacts rapidly with water (reaction 6.27) to produce hydroxyl. When we discuss tropospheric oxidants in §7.1 we shall learn that OH does not last long in the troposphere, but while it exists it can catalyze the destruction of ozone by process (6.24). Ozone is lost by this process at an estimated rate of 1300 Tg y^{-1}.

---

[2]One teragram = 10^{12} g = 10^9 kg, a convenient unit for discussions of global amounts.

Ozone can be created in the troposphere by reactions involving nitric oxide (NO) and hydrocarbons in a manner similar to its formation in urban smog, a subject that we shall discuss in §7.1.2. The creation rate has been estimated at 1300 Tg y^{-1}.

To summarize, we have identified two sources and two sinks, all four of the same magnitude. The tropospheric balance could involve two, three, or all four processes; only better observations will lead to a resolution of the problem.

6.7 Nonhomogeneous reactions

Up to this point we have discussed *homogeneous reactions* in the gas phase. *Nonhomogeneous reactions* involve the presence of surfaces. Surface reactions depend upon the nature and condition of the surface in addition to the properties of the reacting species and introduce a large element of uncertainty into atmospheric chemistry.

For the atmosphere itself, setting aside for the moment the complex surface of the earth, the two areas of nonhomogeneous chemistry that are most important are reactions involving stratospheric aerosols and reactions involving liquid cloud droplets. Homogeneous reactions may also take place in solution inside droplets; raining out is an important means of removing species from the atmosphere.

The discovery of the Antarctic *ozone hole* has provided a spectacular and unexpected example of aerosol chemistry. The aerosols concerned form tenuous sheets at about 20 km, known as *polar stratospheric clouds*. At 20 km in polar regions, the air temperature falls to abnormally low values in winter and early spring. If it falls to about 185 K, ice saturation can occur and ice crystals can form. In the presence of small amounts of nitric acid, crystals may form at temperatures about 10 K higher.

Figures 6.13(a) and (b) show some of the evidence for the Antarctic ozone hole that has developed since 1960. Figure 6.13(a) shows a precipitous decline in the column amount of ozone in October. Figure 6.13(b) shows where the problem lies: Ozone almost vanishes in the spring in the height range 13 to 21 km, where normally it is most abundant.

Springtime in the Antarctic stratosphere has three important characteristics: It is very cold, so that polar stratospheric clouds can form; it is surrounded by an intense vortex that acts as a barrier and retains stratospheric gases within it; and, most obviously, the sun is making its appearance after 6 mo of darkness. The sequence of chemical events that follows from these conditions was demonstrated by in situ measurements from aircraft flying through the wall of the vortex during the spring. Ozone decreased rapidly as the vortex was entered. Inside the vortex the ozone concentration was

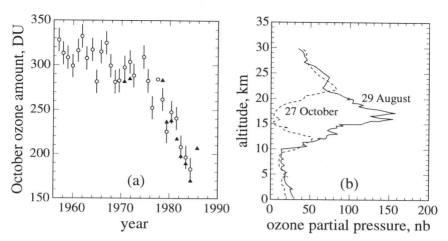

Figure 6.13 The Antarctic ozone hole. (a) Mean ozone column amounts at Halley Bay, Antarctica for the month of October. Some measurements are duplicated. (b) Vertical profiles of ozone measured at McMurdo Sound, Antarctica during winter and spring, 1987.

low, as was also the concentration of NO_2; on the other hand, the concentration of halogens was abnormally high. In the wall of the vortex there were rapid excursions on top of the general trend; in these excursions the same negative correlation between ozone and halogen concentrations was observed.

These data suggest that the abnormal destruction of ozone is associated with catalysis by abnormal levels of active halogens. Chlorine is advected into the Antarctic from other latitudes. It comes mainly in the form of the stable species chlorine nitrate, $ClONO_2$ (see equation 6.35) and as HCl. In order to produce active chlorine in large amounts, chlorine oxide, ClO, must be produced from the chlorine nitrate and NO_2 must be present in such low concentrations that reaction (6.35) does not take place. The release of ClO by itself does not solve the riddle, because ClO cannot be photolyzed by sunlight at 20 km. However, if it exists in very large concentrations it can form the dimer $(ClO)_2$ or Cl_2O_2, which can be photolyzed. The sequence of reactions appears to be

$$\begin{array}{rrcl}
 & ClO + ClO + M & \to & Cl_2O_2 + M \\
 & Cl_2O_2 + \gamma & \to & ClOO + Cl \\
 & ClOO + M & \to & Cl + O_2 + M \\
 & 2(Cl + O_3 & \to & ClO + O_2) \\
\hline
net & 2O_3 + \gamma & \to & 3O_2 \,.
\end{array} \qquad (6.39)$$

Reactions (6.39) are thought to account for 70% of the ozone destruction, while a similar bromine reaction accounts for the remainder.

This brings us to the polar stratospheric clouds, which provide both the means for releasing chlorine from chlorine nitrate and the means for removing active nitrogen species. The following reactions are possible:

$$\text{ClONO}_2(s) + \text{HCl}(s) \rightarrow \text{Cl}_2(g) + \text{HNO}_3(s) \tag{6.40}$$

and

$$\text{ClONO}_2(s) + \text{H}_2\text{O}(s) \rightarrow \text{HOCl}(g) + \text{HNO}_3(s) \, . \tag{6.41}$$

(s) and (g) denote solid and gas phases, respectively. Both of these reactions produce chlorine in the gas phase and leave the nitrogen in the solid phase in the form of nitric acid. The solution of nitric acid in water helps with the formation of stratospheric clouds, the particles of which ultimately settle to lower levels, removing nitrogen from the region of ozone destruction.

6.8 Reading

The two most important books have already been mentioned in Chapter 1. Brasseur and Solomon (1984) discuss the middle atmosphere. Warneck (1988) emphasizes the troposphere, clouds, and aerosols; he offers a "meteorological" point of view.

The December 1989 issue of *Planetary and Space Science* **37**, pp. 1485–1672, commemorates the life's work of G.M.B. Dobson in the field of atmospheric ozone and covers all aspects of the subject in 11 articles. The following two articles give details about the Antarctic ozone hole:

Anderson, J.G., Toohey, D.W., and Brune, W.H., 1991, "Free radicals within the Antarctic vortex: The role of CFC's in the Antarctic ozone loss," *Science* **251**, 39–45,

Schoeberl, M.R., and Hartmann, D.L., 1991, "The dynamics of the stratospheric polar vortex and its relation to springtime ozone depletions," *Science* **251**, 46–52.

6.9 Problems

6.1 Ozone and skin damage. Figure 6.14 shows, in relative units, the damage caused to the skin by an ultraviolet photon. Table 6.3 shows data for a part of the spectral range. The second column is the spectral irradiance above the atmosphere, expressed as the number of incident solar photons per unit area per unit time in each spectral subinterval. Figure 6.13(b) gives ozone densities under normal and depleted conditions. Under depleted conditions the total number of ozone molecules is halved.

OZONE

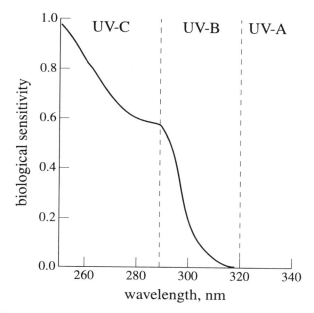

Figure 6.14 Skin damage by sunburn, per photon, in arbitrary units.

Table 6.3 Spectral data

Wavelength, nm	Photon flux, $m^{-2}s^{-1}$	s_R, m^{-2}	$k(O_3)$, m^{-2}
289.8 – 294.1	3.48×10^{18}	6.36×10^{-30}	1.04×10^{-22}
294.1 – 298.5	3.40×10^{18}	5.97×10^{-30}	5.85×10^{-23}
298.5 – 303.0	3.22×10^{18}	5.59×10^{-30}	3.16×10^{-23}
303.0 – 307.7	4.23×10^{18}	5.24×10^{-30}	1.66×10^{-23}
307.7 – 312.5	4.95×10^{18}	4.90×10^{-30}	8.67×10^{-24}
312.5 – 317.5	5.44×10^{18}	4.58×10^{-30}	4.33×10^{-24}
317.5 – 322.5	5.93×10^{18}	4.28×10^{-30}	2.09×10^{-24}

(i) For each subrange in Table 6.3 calculate a relative hazard index, based on the product of the number of photons reaching the surface per unit area with the relative biological sensitivity, taken from Figure 5.14. Add the subranges for a total hazard index. Perform the calculation for normal and depleted conditions using the following data: column abundance of air molecules $= 2.15 \times 10^{29}$ m^{-2}; column abundance of ozone molecules $= 8 \times 10^{22}$ m^{-2} (normal conditions); solar zenith angle $= 60°$.

(ii) The terms UV-A, UV-B, and UV-C are used as indicated in Figure 6.14. Why do you suppose that the medical community makes these distinctions?

6.2 Chemical time constants. Equation (6.16) has the form,

$$\frac{d[O_3]}{dt} = P_{O_3} - D_{O_3},$$

where the production and destruction terms, P_{O_3} and D_{O_3}, are defined in connection with equation (6.36). Consistent with these equations is an equilibrium state, (e), for a motionless atmosphere, for which production and destruction balance,

$$P_{O_3}^e = D_{O_3}^e.$$

Following the derivation of (6.37), and expanding changes in P_{O_3} and D_{O_3} to first order in a Taylor expansion, show that

$$\frac{\partial [O_3]}{\partial t} + \nabla \cdot \vec{F}_{O_3} = -\frac{1}{\tau_c}\left([O_3] - [O_3]_e\right),$$

where

$$\tau_c = -\left\{\frac{\partial (P_{O_3} - D_{O_3})}{\partial [O_3]}\right\}_e^{-1}$$

is the overall *chemical time constant*. $\vec{F}_{O_3}$ is the flux of ozone molecules.

6.3 Lindzen's box model. In Figure 6.15 the left-hand boxes represent low latitudes and the right-hand boxes represent high latitudes. The upper boxes are primarily stratospheric and the lower boxes tropospheric. v and w are meridional and vertical winds. Primed densities are for ozone and unprimed for air. It is assumed that $\rho' \ll \rho$. There is a steady state, $\frac{\partial}{\partial t} \equiv 0$.

(i) From the equation for mass continuity, (F.3), show that

$$w_{4,1}L_{4,1}\rho_4 = v_{1,2}H_{1,2}\rho_1 = w_{2,3}L_{2,3}\rho_2 = v_{3,4}H_{3,4}\rho_3 = C,$$

where C is the mass circulation of air.

(ii) If F_{O_3} is the mass flux of ozone, use Gauss' theorem, (F.19), to show that the volume integral is

$$\int_1 \nabla \cdot \vec{F}_{O_3}\, dV = v_{1,2}H_{1,2}\rho_1' - w_{4,1}L_{4,1}\rho_4' = F(m_1 - m_4),$$

where m in the mass mixing ratio of ozone.

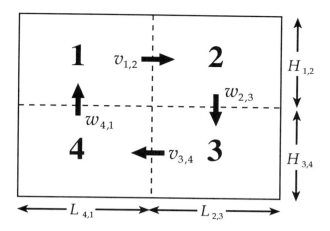

Figure 6.15 Lindzen's model. The densities and velocities in each box are constant. The third dimension is unity.

(iii) From the result of Problem 6.2, show that

$$m_j \left(\frac{1}{\tau_{d,j}} + \frac{1}{\tau_{c,j}} \right) = \frac{m_{j-1}}{\tau_{d,j}} + \frac{m_{j,e}}{\tau_{c,j}},$$

where $j = 1, 2, 3, 4$, $\tau_{c,j}$ is the chemical time constant and $\tau_{d,j}$ is the time taken to empty or to fill box j.

6.4 **Ozone and atmospheric motions.** Use the results of the previous problem to discuss the following.

(i) Assuming the m_j does not differ by orders of magnitude from box to box, discuss the two situations (a) $\tau_{c,j} \gg \tau_{d,j}$ for all j, and (b) $\tau_{c,j} \ll \tau_{d,j}$ for all j. How do your conclusions depend upon the strength of the circulation and the dimensions of the boxes?

(ii) The left-hand boxes are tropical while the right-hand boxes are in darkness during the winter. The division between upper and lower boxes lies at about 20 km. At all times the two lower boxes have chemical time constants long compared with their dynamical time constants (see Figure 6.7). What are the mixing ratios in all four boxes (a) for the summer hemisphere when both of the upper boxes are in photochemical equilibrium, and (b) for the polar night when $\tau_{c,2} \gg \tau_{d,2}$?

(iii) If $\tau_d^{-1} = 2.5 \times 10^{-7}$ s^{-1} in all four boxes and all other data are given in Table 6.4, what are the mixing ratios in each box?

Table 6.4 Data for the box model

Box	m_e	τ_c^{-1}, s^{-1}
1	8×10^{-6}	$\gg \tau_{d,1}^{-1}$
2	3×10^{-6}	2×10^{-7}
3	1×10^{-6}	1.5×10^{-8}
4	3.5×10^{-6}	2.5×10^{-8}

6.5 The Chapman reactions. In this problem we consider only the Chapman reactions, but we ask about the state of excitation of the oxygen atoms. $[O_2]$ at 45 km is given in Table 6.2. Assume $[M] = 5[O_2]$. The temperature is 250 K.

(i) Calculate k_2 and k_3 from Table 6.1 at a temperature of 250 K and estimate J_2 and J_3 at 45 km from Figures 6.3 and 6.4. Hence calculate the equilibrium values of $[O]$ and $[O_3]$, and compare them with the values given in Table 6.2.

(ii) All of the oxygen atoms formed in reaction (6.10) are in the ^3P state, but four fifths of those formed in reaction (6.11) are in the ^1D state. The $O(^1D)$ atoms are thermalized in collisions according to the reaction

$$O(^1D) + M \to O(^3P) + M, \quad k \approx 4.9 \times 10^{-17} \, \text{m}^3\text{s}^{-1}.$$

Write a balanced equation for the production and loss of $O(^1D)$. Assuming the k_2 and k_3 are the same for both forms of atomic oxygen, which is the largest loss term? What is the equilibrium density of $O(^1D)$?

6.6 Odd-nitrogen catalysis. *Note that the equations used here are incomplete, just as were the Chapman reactions in the previous question. However, they do illustrate the nature of the catalytic cycle.* We add the two odd-nitrogen reactions of equation (6.31) to the Chapman reactions and calculate how they decrease the equilibrium amount of ozone for a given concentration of nitric oxide, $[NO] = 5 \times 10^{14}$ m^{-3} at 45 km. Use the same rates and rate constants that you used in the previous problem. In addition, the reaction constant between NO and O_3 in equation (6.31) is 7.0×10^{-21} m^{3}s^{-1}.

Follow the steps that led to equation (6.20) and calculate $[O_3]_e$.

6.7 Odd-chlorine reactions. We use the Chapman reactions and add to them the chlorine reactions (6.32). Again, this is not a complete set of reactions, but from them we may, for a steady state, calculate the concentration of chlorine atoms, [Cl], given the concentration $[Cl_x] = [Cl] + [ClO] = 1.5 \times 10^{14}$ m^{-3} at 45 km. It is not possible to assume that the ozone and atomic oxygen concentrations are unchanged. The following rate constants apply to the conditions at 45 km:

$$k_{Cl,O_3} = 2.7 \times 10^{-17} \text{ m}^3\text{s}^{-1},$$
$$k_{ClO,O} = 4.6 \times 10^{-17} \text{ m}^3\text{s}^{-1}.$$

6.8 **The Antarctic ozone hole.** We consider conditions at 16 km in the Antarctic. This is the level at which the ozone almost disappears in the spring (see Figure 6.13). The rate of disappearance is such that half of an initial ozone concentration of 5×10^{18} m^{-3} is lost in about 30 days. The reactions (6.39) are thought to be responsible. Of these, the slowest (the *rate limiting step*) is the first, the reaction involving two ClO molecules. This statement is true only during the day, when the sun shines. For the other approximately 50% of the time, the cycle simply does not operate.

(i) Show that

$$\frac{d[O_3]}{dt} = -k_{ClO,ClO}[ClO]^2[M].$$

(ii) The reaction rate at the stratospheric temperature of 190 K is 8×10^{-44} m^{6}s^{-1}. Calculate the concentration of ClO that could be responsible for the observed loss. $[M] = 3 \times 10^{24}$ m^{-3} at 16 km.

CHAPTER 7

TOPICS IN TROPOSPHERIC CHEMISTRY

Three related topics illustrate important principles of tropospheric chemistry. First, we discuss tropospheric oxidants. Oxidants control the scavenging of many harmful atmospheric pollutants, but they can also be a part of the pollution problem by generating ozone and other species that irritate living tissues. The second and third topics, sulfur and carbon, are examples of species controlled by huge reservoirs at the surface of the earth, including rocks, the oceans, and the biosphere. Sulfur forms reactive species that ultimately affect the stratosphere, creating a link between the state of the surface and the state of the middle atmosphere. Carbon is less reactive; interest lies in its distinctive reservoirs, the most complex being the biosphere, and the cycles of carbon between them.

The most important oxidants are hydroxyl, atomic oxygen, and the peroxy radicals. These, together with ozone, can be formed in chain reactions when sunlight falls on a mixture of nitrogen oxides and hydrocarbons, such as cars emit copiously. Photolysis of a trace of natural ozone releases the hydroxyl that starts the chain reaction leading to the many dangerous constituents that make up urban smog.

Sulfur and carbon have both similarities and differences. Both elements are involved in the life cycles of plants and animals, as a result of which they are brought to the atmosphere in reduced forms. More commonly, both elements occur as oxidized species: carbon dioxide, sulfur dioxide, and sulfates, including sulfuric acid. Both elements have similar geological cycles for the oxidized species, involving weathering of rocks, sedimentation in the oceans, followed by subduction that creates land sediments for further weathering. For the most part, these cycles are closed, with the important exception of volcanism, which can place oxidized species high in the atmosphere. A further similarity involves industrial effluents. Oxidized species of both elements are emitted copiously from the combustion of fossil fuels. Sulfur emissions may lead to acid rain. Carbon dioxide affects the atmospheric greenhouse. Both raise important environmental issues.

In other matters the two elements differ. Oxidized sulfur leaves the surface in the form of sea spray and industrial effluents. Almost all returns rapidly to the surface, some as acid rain. Reduced species of sulfur emitted by the land and marine biospheres also, for the most part, oxidize and return rapidly to the surface. An important exception is carbonyl sulfide, with a chemical lifetime in the lower atmosphere of more than 40 years. Carbonyl sulfide is evenly mixed through the troposphere and the lower stratosphere. Above most of the ozone, photolysis of carbonyl sulfide is possible with the end point, after oxidation, of sulfuric acid, which forms an aerosol layer near 20 km (the Junge layer). This layer can be reinforced by volcanic activity.

Our discussion of carbon centers on the major reservoirs and the cycles between them. The short-term reservoirs include the atmosphere (in which the carbon is principally in the form of carbon dioxide), the ocean mixed layer, and a part of the land biosphere. These reservoirs can equilibrate in less than 10 y. If all the carbon in them were released to the atmosphere, the concentration of carbon dioxide would approximately double over its present level.

Very much larger quantities of carbon exist in intermediate reservoirs, which can fill the atmosphere in times between 100 and 1000 y. One of these reservoirs is fossil fuel, which is responsible for the increase of atmospheric carbon dioxide observed in recent decades. Finally, huge reservoirs exist in the continental and marine sedimentary deposits. The sedimentary cycle is of scientific interest but is not important for current environmental concerns. Volcanism is the main link between the sedimentary cycle and the atmosphere. Volcanic gases are rich in carbon dioxide.

7.1 Oxidants

7.1.1 The free radicals

An interrelated group of free radicals controls the oxidation state of the lower atmosphere. These are, principally, $O(^3P)$, $O(^1D)$, OH, HO_2, and RO_2, where R stands for a hydrocarbon radical derived from the stable compound RH. RO_2 and HO_2 are called *peroxy free radicals*. All free radicals are present in the troposphere only in very low concentrations. OH averages about 5×10^{11} molecules m^{-3}, while $O(^3P)$ averages about 1×10^{10} molecules m^{-3} (compare to 2.7×10^{25} molecules m^{-3} for dry air at stp). The oxidants are subject to large changes in concentration if ambient chemical conditions change or if the insolation changes; the ability of the atmosphere to tolerate certain anthropogenic gases is hostage to such possibilities.

Of the tropospheric oxidants, OH is the most important. It does not react with the major atmospheric constituents, O_2, N_2, H_2O, CO_2 and Ar, but reacts readily with all important minor species, CH_4, CO, H_2, O_3, NO_2, SO_2, NH_3, HCHO, etc. In the body of the troposphere most reactions are with CH_4 and CO, but reactions with NO_2 and HCHO are more important in the boundary layer. The lifetime of OH for these reactions lies between 0.3 and 2.5 s. Most reactions with OH generate new free radicals, from which OH may be regenerated. These catalytic chains are terminated when OH reacts with NO_2 or with another free radical.

7.1.2 Smog-forming reactions

We may trace a high level of interest in tropospheric oxidants to the formation of smog in the Los Angeles basin. This subject offers a convenient and interesting way to enter a complex set of ideas, some of which are also relevant to the nonpolluted atmosphere.

The essential ingredients for smog formation are nitrogen oxides and volatile organic compounds, mixed in the presence of ultraviolet radiation. Nitric oxide and volatile organics are common products of human activities, specifically and copiously from incomplete combustion in automobile engines. Figure 7.1 shows a typical sequence of chemical changes during a smog event in Los Angeles. Nitric oxide and hydrocarbons begin to build up at the start of the morning rush hour. During the course of the day deleterious products appear, represented here by ozone, aldehydes, and peroxyacetyl nitrate (PAN). All three can irritate the tissues of the eye and nose, can harm green plants, and can even cause physical damage to paint surfaces. Of these irritants, we shall restrict our discussion to ozone. The formation of ozone involves OH, and OH can, in other reactions, lead to numerous other products, including aldehydes and PAN.

If we ask only about ozone formation, the problem may be broken down into a series of questions, partly circular. Given that ozone formation follows from the reaction between molecular and atomic oxygen, equation (6.12), the first question is:

Where does the atomic oxygen come from? The problem has been briefly discussed in §6.6. The photolytic source, equation (6.10), is unimportant in the troposphere because the required actinic radiation does not penetrate so far down. However, the photolysis of nitrogen dioxide (NO_2) can also yield atomic oxygen, and this may be accomplished by the absorption of blue radiation, which penetrates with little attenuation to the earth's surface:

$$NO_2 + \gamma \to NO + O(^3P), \quad \lambda < 400\ nm, \tag{7.1}$$

followed by reaction (6.12). This leads to the next question:

Where does the nitrogen dioxide come from? Nitric oxide (NO) and hydrocarbons (RH, where R is a hydrocarbon radical) are available from the

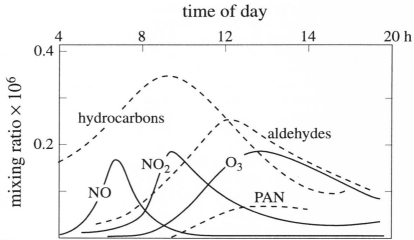

Figure 7.1 Chemical changes during a Los Angeles smog event. This was a day of intense eye irritation.

incomplete combustion of fossil fuels (see Figure 7.1). In the presence of the OH free radical the following reactions are possible:

$$RH + OH \to R + H_2O , \qquad (7.2)$$
$$R + O_2 + M \to RO_2 + M , \qquad (7.3)$$
$$RO_2 + NO \to NO_2 + RO . \qquad (7.4)$$

Each of the above reactions is an example of the destruction of a free radical followed by the formation of another free radical: OH, R, and RO_2 destroyed; R, RO_2, and RO formed. Nitric oxide and hydrocarbons also exist in the nonpolluted atmosphere, but in very low concentrations, and the above equations are also relevant to the formation of tropospheric ozone in a clean atmosphere. Finally, to complete the chain of questions:

Where does the hydroxyl radical come from? This is where the argument becomes circular. The largest sources of hydroxyl involve reactions of water and methane with the reactive form of atomic oxygen, $O(^1D)$ (see equations 6.27 and 6.28). The most likely source of $O(^1D)$ is the photolysis of ozone, equation (6.11). Radiation in the Huggins bands penetrates into the troposphere and is energetic enough to create excited species of both molecular and atomic oxygen:

$$O_3 + \gamma \to O_2(^1\Delta_g) + O(^1D) . \qquad (7.5)$$

According to equation (6.27) this sequence can result in the production of two reactive OH radicals from one ozone molecule. On the other hand, reactions (7.2), (7.3), and (7.4) form one ozone molecule with the destruction of only one hydroxyl. The combination of the two sequences forms a catalytic chain reaction that can, given the right conditions, generate important amounts of hydroxyl from an initial trace of tropospheric ozone.

7.1.3 Other reactions with oxidants

The above reactions are illustrative of a class, others of which lead to similar results under smog conditions. Two alternatives to (7.2), (7.3), and (7.4) involve atmospheric formaldehyde (HCHO) or methane (CH_4). The formaldehyde reactions that lead to the formation of NO_2 are

$$HCHO + OH \rightarrow HCO + H_2O , \tag{7.6}$$

$$HCO + O_2 \rightarrow HO_2 + CO , \tag{7.7}$$

$$HO_2 + NO \rightarrow NO_2 + OH . \tag{7.8}$$

Equation (7.7) introduces another oxidant, the hydroperoxy radical HO_2.

Methane (CH_4) is the most abundant hydrocarbon in the atmosphere and is second only to carbon dioxide as a source of carbon. It is emitted to the atmosphere as a result of bacterial degradation of organic matter, primarily by ruminants, but also by microbial decay in anaerobic sludges in lakes, rice paddies, and swamps. The concentration of methane in the atmosphere is increasing with time, presumably because of increasing human activity. Before 1700, the methane concentration, as measured in air incorporated as bubbles in major ice sheets, was 0.7 parts per million by volume (ppmv, see Appendix A); now it is close to 1.7 ppmv and is increasing by 0.02 ppmv annually.

The major tropospheric sink for methane (and the major source of formaldehyde) is the reaction between methane and OH:

$$OH + CH_4 \rightarrow H_2O + CH_3 , \tag{7.9}$$

$$CH_3 + O_2 + M \rightarrow CH_3O_2 + M , \tag{7.10}$$

$$CH_3O_2 + NO \rightarrow CH_3O + NO_2 , \tag{7.11}$$

$$CH_3O + O_2 \rightarrow HCHO + HO_2 . \tag{7.12}$$

Equations (7.9) to (7.12) consume OH but create HO_2, from which OH may be recovered by reaction (7.8). They also produce formaldehyde, which can fuel the sequence (7.6), (7.7), and (7.8). All of the oxidant cycles that we have discussed so far are, in fact, linked together, with consequences that are difficult to describe in simple terms.

Cycles cease with a chain-limiting step, which is particularly effective if it is a reaction between two oxidants. Reaction (6.13), involving two odd-oxygen species, is an example that we have already encountered. Others that involve HO_2 are

$$HO_2 + HO_2 \rightarrow H_2O_2 + O_2, \quad (7.13)$$
$$HO_2 + CH_3O_2 \rightarrow CH_3OOH + O_2, \quad (7.14)$$
$$HO_2 + OH \rightarrow H_2O + O_2. \quad (7.15)$$

All of the products in reactions (7.13), (7.14), and (7.15) are stable or relatively stable species.

7.2 Sulfur

The concentration of sulfur in the free atmosphere is exceedingly small. Distant from local sources, the most abundant sulfur species is carbonyl sulfide (COS), which is present in a concentration of 0.5 parts per billion by volume (ppbv, see Appendix A). Nevertheless, the cycles of atmospheric sulfur are important for the following reasons amongst others: Sulfur is a component of certain amino acids that are essential to the functioning of life as we know it; the end point of sulfur oxidation is sulfuric acid (H_2SO_4) and the sulfate ion (SO_4^{--}), both of which can influence water droplet growth and cloud formation (see Chapter 8) and, indirectly, the climate; and sulfuric acid is the noxious component of *acid rain*, a well-known environmental hazard. Sulfur has major anthropogenic sources, and provides a route for human intervention in the environment. Sulfur is interesting because of the interactions between different classes of process—biological, geological, and atmospheric—each with very different cycling rates.

7.2.1 Sources and sinks

All sources of atmospheric sulfur are at the surface. They fall into two distinct groups, differing in their state of oxidation. A *reduced group* includes hydrogen sulfide (H_2S), carbonyl sulfide (COS), dimethyl sulfide or DMS (CH_3SCH_3), and carbon disulfide (CS_2); this group is biogenic in origin and cannot have been long in the atmosphere or they would be oxidized. The second group is *oxidized* because it has been exposed to oxygen, either at high temperatures or for a long time. The two most important oxidized sources of sulfur are sulfur dioxide (SO_2) and sulfate ions (SO_4^{--}); these sources can be either geochemical or industrial.

We first consider the oxidized sources. The geochemical cycle is illustrated in Figure 7.2. This figure applies generally to all species that cycle between the surface rocks, the atmosphere, and the oceans, and not to sulfur alone; it is, for example, also relevant to the carbon cycle (§7.3). The bottom part of the figure shows processes in the crust and mantle, which are

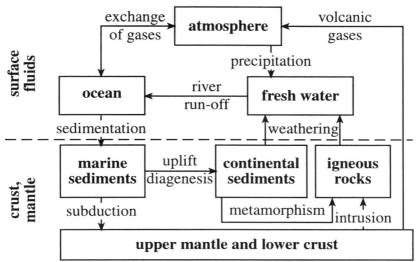

Figure 7.2 The geochemical cycles.

exceedingly slow. These are included for the sake of completeness, but unless we wish to discuss "geological" time scales ($\sim 10^8$ y or longer), we may regard the crust as static. The upper part of Figure 7.2 shows processes that cycle rapidly compared with those in the solid earth. The connecting processes between the two groups are weathering of surface rocks (involving the action of water and dissolved CO_2), volcanism, and sedimentation (the sink).

The surface rocks contain huge amounts of sulfur compared with the sulfur in the atmosphere and oceans, and the flux of sulfur from weathering is large compared with all other fluxes that we shall discuss. Nevertheless, as far as we can tell, the weathering flux is matched by precisely equal amounts of sedimentation, with no net effect on the atmosphere. Weathering involves the disintegration of sulfate-containing rocks, followed by the transport of weathered material into the oceans. Sulfates from the oceans can enter the atmosphere in the form of sea spray, but measurements indicate that this spray simply returns to the oceans without interacting with the atmosphere. Much smaller quantities of sulfur compounds reach the surface as the result of oxidation of reduced sulfur compounds, or as the end point of volcanic processes. These sulfur compounds are known as *excess sulfate* and are the most interesting part of the atmospheric sulfur cycle.

The volcanic source is mainly SO_2 and has the distinctive property of conveying some material directly from the surface to the stratosphere. We

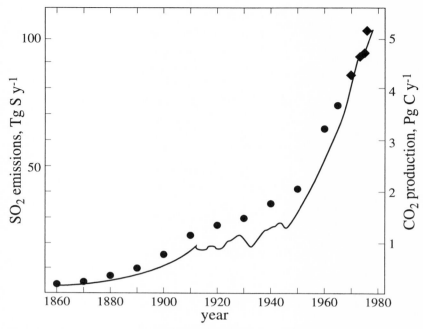

Figure 7.3 Anthropogenic emission of SO_2 and CO_2. Solid curve, CO_2. Filled circles and diamonds, SO_2. Tg, teragram = 10^9 kg; Pg, petagram = 10^{12} kg.

shall examine the consequences of this in §7.2.3. Near the surface, the main source of oxidized sulfur is human activity, particularly the combustion of fossil fuels. This source is similar to the volcanic source in that crustal materials are oxidized at high temperatures and injected into the atmosphere. It differs in being larger and in placing the sulfur in the boundary layer rather than in the stratosphere. The emission of anthropogenic sulfur closely parallels the emission of anthropogenic carbon dioxide (see Figure 7.3), and both are increasing as industrial activity increases.

Sources of reduced biogenic sulfur differ over land and sea. Over land, the main emissions are H_2S and DMS, followed by CS_2. Some 10% of land emission is COS, which, being more stable and having a longer lifetime than other reduced compounds, is important far from the main sources (§7.2.3). Over the oceans the main biogenic source is DMS.

All sulfur sources are variable both in space and in time. Total amounts summed over the entire atmosphere are less variable; it is usual to give global totals in units of TgS y^{-1} (teragrams of sulfur per year, where 1 teragram is 10^9 kg or 1 megaton). Under this convention only the element sulfur is accounted for, regardless of its state of combination; a similar convention is used for carbon in §7.3.

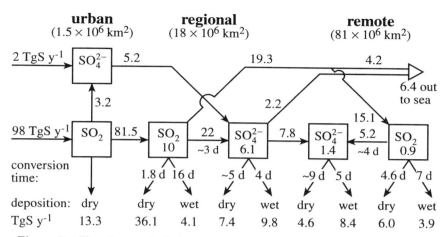

Figure 7.4 Flux diagram for industrial sulfur emissions. The diagram applies to continental land masses in the northern hemisphere (total area 3×10^8 km^2). The numbers beside arrows are fluxes in TgS y^{-1}. Numbers in boxes are the total amounts of sulfur integrated over a vertical column in units of mgS m^{-2}. Conversion times are in days.

The oxidized sources are 7 TgS y^{-1} for volcanism, 98 TgS y^{-1} for industrial SO$_2$, and 2 TgS y^{-1} for industrial sulfate. The reduced sources are 36 TgS y^{-1} for the oceans and 7 TgS y^{-1} for the land. The total source of sulfur is 150 TgS y^{-1}. All of this sulfur returns to the ground either in solution in water droplets (*wet deposition*) or directly from the gas phase (*dry deposition*). This deposition is oxidized, but not always as far as sulfate. It is estimated that this deposition consists of 86 TgS y^{-1} in the form of SO$_2$ and 70 TgS y^{-1} in the form of excess sulfate. In the next two sections we consider the reactions that take place while the sulfur moves from sources to sinks.

7.2.2 Reactions in the boundary layer

Apart from COS, very little sulfur from surface sources leaves the boundary layer. Figure 7.4 shows, in general terms, how anthropogenic (oxidized) sulfur passes from sources to sinks (deposition). Of the total sulfur emitted, 13% is deposited in urban areas, where it originated, 57% is deposited in larger areas that are known to be directly impacted by industrial emissions, for example, central Europe or the northeastern United States, and the remaining 30% is spread widely. At each stage in its journey SO$_2$ can be oxidized to SO$_3$, which combines with water to form sulfuric acid and deposits as acid rain.

The manner in which SO$_2$ is oxidized is debatable, and a number of pathways have been proposed. The most straightforward involves the excitation of SO_2 by the absorption of a solar quantum in the visible spectrum, providing enough energy to overcome the activation energy of the reaction with molecular oxygen:

$$SO_2 + \gamma \rightarrow SO_2^*, \qquad (7.16)$$

$$SO_2^* + O_2 \rightarrow SO_3 + O, \qquad (7.17)$$

$$SO_3 + H_2O \rightarrow H_2SO_4. \qquad (7.18)$$

Collisions with nitrogen molecules can compete with (7.17), thermalizing the excited SO_2 molecule and stopping the reaction.

More important is a second group of reactions that involves tropospheric oxidants. One example is

$$SO_2 + OH + M \rightarrow HOSO_2 + M, \qquad (7.19)$$

$$HOSO_2 + O_2 \rightarrow SO_3 + HO_2. \qquad (7.20)$$

Based on an OH concentration of 10^{15} m^{-3}, these two reactions together lead to a lifetime for SO_2 of ∼10 d, short enough for polluted air crossing a continent to be oxidized to sulfuric acid, at least in part.

This is an appropriate point to bring in the possibility of the oxidation of SO_2 to sulfuric acid in liquid drops. SO_2 dissolves in water to form sulfurous acid (H_2SO_3), which ionizes to HSO_3^- and SO_3^{--}. These ions may be oxidized by oxidants that are in solution, particularly by H_2O_2. Schematically,

$$SO_3^{--} + \tfrac{1}{2}O_2 \rightarrow SO_4^{--}. \qquad (7.21)$$

Reaction (7.21) only proceeds rapidly in the presence of catalysts, of which ferrous and manganese ions appear to be the most important. Natural water contains these metallic ions in sufficient quantities to act as catalysts; these aqueous reactions in droplets may be a more important source of sulfuric acid than the gas phase reactions, especially at night.

Turning now to reduced sulfur compounds, a flux diagram for the processes in the marine boundary layer is shown in Figure 7.5. Here, the most abundant reduced sulfur compound is dimethylsulfide (DMS), which can react with OH in the gas phase in the presence of molecular oxygen, although details of the reaction are not understood. The products are 25% SO_2 and 50% methanesulfonic acid (MSA, CH_3SO_3H); the remaining 25% is indicated as "unknown" in Figure 7.5.

The presence of methanesulfonic acid in marine aerosol has attracted attention because it is the most abundant sulfur compound not attributable to sea spray. The occurrence of significant proportions of MSA in marine aerosol has led to the speculation that this chemical may be involved in the condensation of droplets in the boundary layer. It is possible for these droplets to grow to form boundary-layer clouds and for these clouds to

TOPICS IN TROPOSPHERIC CHEMISTRY

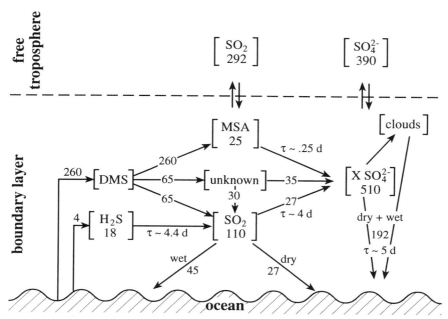

Figure 7.5 Flux diagram for sulfur in the marine boundary layer. Fluxes are in units of $\mu gS\ m^{-2}\ d^{-1}$ (100 $\mu gS\ m^{-2}d^{-1}$ = 13.2 TgS y^{-1} when summed over all the oceans). Numbers in square brackets are column amounts in $\mu gS\ m^{-2}$, evaluated independently for the boundary layer and the atmosphere above. COS is not included but is discussed elsewhere. DMS = dimethylsulfide, MSA = methanesulfonic acid. τ = conversion time.

increase the albedo of the planet, possibly changing the climate, thereby creating a link between the climate and life in the oceans — a small argument in favor of Gaia.

7.2.3 Reactions in the troposphere and stratosphere

Despite the mixed units in Figures 7.4, 7.5, and 7.6,[1] it is possible to conclude from them that the sulfate from surface sources is largely restricted to the boundary layer, and that SO_2 and SO_4^{--} have relatively constant concentrations outside the boundary layer. The amount of sulfur in the troposphere is approximately the same for the two species. For SO_2 the

---

[1] There is some excuse for this lack of consistency. For gaseous species, the molecular mixing ratio and parts by volume are sensible units, particularly for a well-mixed atmosphere. Sulfate radicals, however, are usually present in solid or liquid aggregates. For such, densities in mass per unit volume at stp (this is really a mass mixing ratio, i.e., mass per 1.293 kg of air) are more appropriate. If the sulfur were evaporated from the condensed phase to form a monatomic gas, the equivalence between the two systems is that 0.03 $kgS\ m^{-3}$ corresponds to 0.07 ppbv.

mixing ratio is 0.07 ppbv, 10^{-2} to 10^{-3} times the concentration in polluted areas. COS is the least variable sulfur species in the atmosphere, with a mixing ratio throughout the boundary layer and troposphere of 0.5 ppbv.

This difference in the behaviors of sulfur species is best understood in terms of relative lifetimes. Reduced sulfur compounds, with the exception of COS, have lifetimes with respect to tropospheric oxidants of days or fractions of a day. SO_2 is oxidized to sulfate in a few days and is leached out of the atmosphere in another few days. These times are short compared with the time to mix between distant parts of the troposphere or between troposphere and stratosphere (both ~ 1 y): As we discussed with respect to Figure 7.4, sulfur pollution is a local or regional problem. By way of contrast, COS has an estimated lifetime with respect to tropospheric oxidants of 44 y. A constant mixing ratio for COS is to be expected (and is observed) up to levels where photolysis begins or where oxidant concentrations become very high.

Reactions of COS with oxidants are

$$COS + OH \rightarrow CO_2 + HS, \qquad (7.22)$$
$$COS + O \rightarrow CO + SO, \qquad (7.23)$$

followed by rapid reactions leading to SO_2. Although these reactions are too slow to affect the tropospheric concentration of COS, they may be fast enough to offset the slow oxidation of SO_2 to sulfate (equations 7.16 to 7.20). In other words, both the SO_2 and the H_2SO_4 observed in the upper troposphere could be products of the decomposition of COS.

The photochemical dissociation of COS appears to follow:

$$COS + \gamma \rightarrow CO + S(^3P \text{ or } ^1D), \quad \lambda < 250 \text{ nm}, \qquad (7.24)$$
$$S(^3P) + O_2 \rightarrow SO + O, \qquad (7.25)$$
$$SO + O_2 \text{ or } O_3 \rightarrow SO_2 + O \text{ or } O_2. \qquad (7.26)$$

A calculation of COS and SO_2 concentrations based upon equations (7.22) to (7.26) is shown in Figure 7.6.

SO_2 can be photolyzed to $SO + O$ by ultraviolet radiation in the lower stratosphere. However, the net result is to leave the SO_2 concentration unchanged and to generate oxygen atoms by reaction (7.26). The chain-terminating reaction for sulfur is thought to be the reactions (7.19) and (7.20) with sulfuric anhydride (SO_3) as the end product. Some of the sulfuric anhydride combines with water to form sulfuric acid but in the dry air of the stratosphere some will remain in gaseous form, for a short time, as the data in Figure 7.6 indicate.

The end point of this discussion of sulfur in the troposphere and stratosphere is sulfuric acid, principally in droplet form. The data in Figure 7.6 show that sulfuric acid aerosol tends to concentrate in a layer near 20 km,

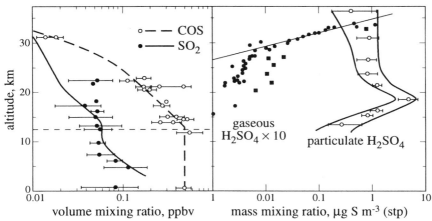

Figure 7.6 Calculated and observed concentrations of sulfur compounds in the troposphere and stratosphere. The calculations based on equations (7.22) to (7.26) are the two curves on the left side. The points are observations. The tropospheric concentration of COS is given, and it is further assumed that COS is decomposed in the stratosphere by photons with wavelengths less than 312 nm. The panel to the right contains observed data on H_2SO_4. The full line is the equilibrium concentration of the gaseous component over a 3:1 mixture of sulfuric acid and water.

the *Junge layer*. The level of maximum density of aerosol at 20 km is below the level of most rapid formation of sulfuric acid and is thought to be a consequence of sedimentation. As particles fall into denser air they sink more slowly, and the concentration must increase in order to maintain a steady-state flux. Just below the tropopause, turbulent mixing suddenly increases (Figure C.4), and the aerosol mixes rapidly down to where it can be leached out by precipitation. The two effects, acting together, are probably responsible for the layered structure, although a quantitative theory has yet to be given.

Direct sampling shows that the stratospheric droplets consist of 75% sulfuric acid and 25% water. At lower levels, the sulfuric acid is usually combined to form a salt. The commonest combination is with ammonia to form ammonium sulfate, $(NH_3)_2SO_4$, and ammonium persulfate, NH_3HSO_4.

Under most circumstances the peak concentration of particles in the Junge layer is insufficient to change the optical properties of the atmosphere in an observable way. However, the layer happens to lie at an altitude that is particularly important for some aspects of the twilight, and it is possible for the aerosol to increase to the degree that its effects are visible to an observer.

When the sky is clear, an observer who looks towards the rising or setting sun, when the sun is 3 to 4° below the horizon, will see a faint, diffuse patch called the *purple light*, centered about 20° above the horizon. At this time, the sun does not directly illuminate that part of the lower atmosphere in the observer's line of sight, and most of the observed light is scattered from a layer 20 to 30 km above the ground, far beyond the horizon. The purple color is caused by the long absorption path through ozone, absorbing yellow and green light in the Chappuis bands (Figure 3.2c). If the aerosol concentration in the Junge layer increases by a factor of 10 over its normal value, it becomes the dominant scatterer at these levels, and the purple light may intensify, sometimes with dramatic visual consequences. Brilliant twilights are a common aftermath of major volcanic eruptions, which can deposit huge amounts of SO_2 into the stratosphere, overwhelming the steady-state processes that we have discussed.

The 1963 explosion of Mt. Agung in Bali is estimated to have injected 12 TgS into the stratosphere, approximately 100 times the annual global supply of COS. After the explosion, the peak concentration of particles in the Junge layer rose by a factor of about 10, and brilliant twilights were observed worldwide. There was a time lag of about 1 y between the eruption and the occurrence of the largest stratospheric aerosol concentrations, demonstrating empirically that the aerosol was formed indirectly from the material injected by the eruption. The Junge layer did not return to normal until 1967.

7.3 Carbon dioxide and the carbon cycle

7.3.1 *A perspective*

Carbon dioxide is chemically inert in the troposphere and stratosphere, and we shall not treat its homogeneous chemistry in this section. Instead, our discussion will focus on the identifiably different ways in which the carbon is stored (in the *reservoirs*) and the *fluxes* of carbon between reservoirs, with a focus on the factors that control the atmospheric reservoir of carbon dioxide.

Carbon dioxide is important for a variety of reasons. It is the main atmospheric reservoir of carbon and is an essential link in the chain of terrestrial life. It dissolves in water to form carbonic acid, bicarbonate, and carbonate ions:

$$\begin{aligned} CO_2(\text{gas}) &\rightleftharpoons CO_2(\text{aqueous}), \\ CO_2 + H_2O &\rightleftharpoons H_2CO_3(\text{carbonic acid}), \\ &\rightleftharpoons H^+ + HCO_3^-(\text{bicarbonate ion}), \\ &\rightleftharpoons 2H^+ + CO_3^{--}(\text{carbonate ion}). \end{aligned} \qquad (7.27)$$

Carbonic acid, while very weak, is the main agent for weathering surface rocks and maintaining the geological cycle. Carbon dioxide is, after water vapor, the second most abundant polyatomic gas in the atmosphere. Strong infrared absorption bands, principally the 15 μm bands, make important contributions to the thermal state of the atmosphere. The increasing concentration of carbon dioxide in the atmosphere due to human activities, and its effect upon the climate, adds a sociopolitical dimension to studies of this important gas.

A discussion of carbon dioxide involves many very different time scales, from hours to billions of years, and this is perhaps the most interesting aspect of current research on carbon dioxide. It is also one of the most confusing. The importance of a time scale is a judgment based on the professional interests of the scientist or upon social considerations. What is crucial for one set of interests may be of no significance for another. Acquiring a sense of perspective is important for any investigator.

If we go back to the origin of the planet, carbon condensed from the proto-planetary nebula about 4.6 By ago in the form of carbon dioxide ice. This was later incorporated into rocks as the planets accreted and subsequently transferred, in part, to the atmosphere by volcanic processes. The earth as a whole has changed little in the past 10^9 y. The time for volcanism to fill the atmospheric reservoir is much less than this, less than a 10^5 y. The storage of carbon in ocean sediments is required if oxygen is to be released to the atmosphere (see §1.4.3); storage in sediments may take a 10^6 y. To go to the other extreme, near the surface of the earth carbon dioxide has a diurnal cycle that is clearly of biospheric origin. The observed build up of carbon dioxide in the atmosphere from human activities involves storage in the biosphere and the oceans, and a variety of time scales from seasons to millennia.

In order to track the pathways of carbon, it is convenient to identify the sizes of independent reservoirs and the sizes of the fluxes of carbon between pairs of reservoirs. The content of the ith reservoir, G_i, is usually given in PgC, petagrams[2] of carbon, regardless of the state of chemical composition of the carbon; fluxes between the ith and jth reservoirs, (F_{ij}), are given in PgC y^{-1}.

The lifetime of the ith reservoir with respect to a flux to the jth reservoir is

$$\tau_{ij} = \frac{G_i}{F_{ij}}. \qquad (7.28)$$

The order of the suffices is important. If the two reservoirs are in detailed balance (as will be the case if the system is close to equilibrium), $F_{ij} = F_{ji}$, and

$$\frac{\tau_{ij}}{G_i} \approx \frac{\tau_{ji}}{G_j}. \qquad (7.29)$$

As was the case for sulfur, carbon reservoirs may be divided in *oxidized*

---

[2] Reminder: Pg = 10^{12} kg or 10^3 Tg.

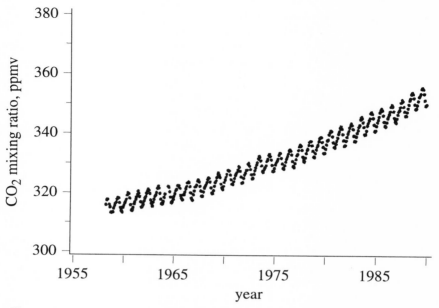

Figure 7.7 Monthly mean carbon dioxide concentrations. Measurements made at the Mauna Loa observatory, Hawaii.

species—CO_2, HCO_3^-, CO_3^{--}—together with the metallic carbonates, principally $CaCO_3$ (limestone), and *reduced species*, consisting of free carbon in the rocks and carbon in the biosphere and methane in the atmosphere. As for sulfur, the reduced species of carbon depend upon the existence of terrestrial life, which is reflected in the alternate terms *inorganic* for oxidized species and *organic* for reduced species.

7.3.2 Reservoirs

Table 7.1 lists the minimum set of reservoirs than are currently used to classify the processes controlling CO_2, listed in order of the times for each to empty or to fill the atmospheric reservoir. The order is also roughly one of increasing reservoir size.

Atmospheric reservoir

Figure 7.7 shows a long record of monthly averaged CO_2 concentrations measured on Mauna Loa, Hawaii. In 1989 the annual mean mixing ratio was 352 ppmv and was approximately the same worldwide and throughout the troposphere and the stratosphere; it corresponded to a global inventory

Table 7.1 The carbon reservoirs

Reservoir	Size, PgC	τ, y	Classification
Atmosphere	7.5×10^2 (i)	—	
Land biosphere (short term)	1.5×10^2 (o)	7	Short term
Ocean mixed layer	6.7×10^2 (i)	7	
Land biosphere (intermediate)	2.1×10^3 (o)	70	
Fossil fuels	7.0×10^3 (o)	125	Intermediate
Deep oceans	3.7×10^4 (i)	1000*	
	1.0×10^3 (o)		
Continental sediments	3.4×10^7 (i)	3×10^4	
	1.0×10^7 (o)		Long term
Marine sediments	3.0×10^7 (i)	$10^8 - 10^9$	
	0.2×10^7 (o)		

τ is the time for the atmospheric reservoir to fill from the reservoir in question. The terms *short*, *intermediate*, and *long* refer to the size of τ. (o) stands for organic; (i) stands for inorganic.
*Turnover time for the deep oceans.

of 773 TgC.

At all observing sites where it is measured, carbon dioxide shows a seasonal variation of concentration. In Hawaii the amplitude of the variation is about ±1%. The amplitude of the seasonal change is larger in the Northern than in the Southern Hemisphere, because most of the active, seasonally varying, land biosphere is in the North.

At the present time, the CO_2 concentration is increasing by about 0.39% per year. Since its discovery, this increase has been attributed to the burning of fossil fuels (see Figure 7.3 for industrial production rates of CO_2). The accepted value of the mixing ratio prior to the start of the industrial revolution is 280 ppmv. Over longer time scales there are natural changes in the mixing ratio. Inclusions in ice cores from ancient ice sheets indicate variations from 175 to 300 ppmv during the last ice age.

The current rate of increase is, however, less than the rate at which anthropogenic carbon enters the atmosphere. The figure 0.39% per year represents 3 TgC y^{-1}, while extrapolation to the present in Figure 7.3 gives an anthropogenic rate of 6 TgC y^{-1}. The difference must be going directly into surface reservoirs.

Biospheric reservoirs

As far as exchange of CO_2 with the atmosphere is concerned, the important biospheric reservoirs are on land. The marine biosphere plays a part in forming oceanic sediments, which allows the carbon to be sequestered or isolated from the atmosphere on geological time scales, but direct communication between the marine biosphere and the atmosphere is small compared with that between the land biosphere and the atmosphere.

For the land biosphere, there is an important distinction to be made between *short-term* and *intermediate storage*. Short-term storage involves animals and leafy structures, either dead (leaf litter) or alive. A reasonable estimate of the lifetime for short-term storage is 1 to 10 y. Intermediate storage involves structural elements of trees (trunk, roots, limbs), with lifetimes of ~ 100 y, and buried litter or humus, which is cycled in ~ 200 y. There is a continuum of behavior between these two extremes, but the general distinction between "short-term" and "intermediate" is relevant to the fate of industrial CO_2. Short-term and intermediate reservoirs are assigned amounts of 150 and 2100 PgC in Table 7.1.

The behavior of the biospheric reservoirs is exceedingly complex, but a few general comments are possible. Animals are not a large factor in the short-term storage; the more important component is the leaves and the litter deposited annually on the earth's surface. CO_2 enters the short-term storage by photosynthetic processes in green plants, for which we may write schematically,

$$n CO_2 + n H_2O + \gamma \to (CH_2O)_n + n O_2 . \tag{7.30}$$

There is also a reverse path. Animals breath, plants respire in the dark, and microbial and herbivore activity oxidizes leaves and litter. These processes all return CO_2 from short-term storage back to the atmosphere.

The growth of trees also utilizes leaf photosynthesis. A return path exists through the breakdown of dead structural material by microbial activity. There is a relatively small contribution from burning.

Oceanic reservoirs

The mean depth of the oceans is 3730 m, of which a small fraction is stirred by winds and can communicate rapidly with the atmosphere; this *mixed layer* is conventionally assigned a mean depth of 75 m. Beneath the mixed layer, between 100 and 1000 m deep, lies the *thermocline*, a region in which temperature and salinity vary with depth. Beneath the thermocline lies the *deep ocean*, about which comparatively little is known, but which is usually assumed to be homogeneous. The simplest chemical models distinguish only between an accessible mixed layer and a sluggish combination of thermocline and deep ocean. The deep ocean and the thermocline are assumed, with exceptions, to communicate with the atmosphere through the mixed layer. The exceptions are limited sinking regions in both polar regions, which allow a direct connection between the atmosphere and the deep ocean. The average turnover time for the deep oceans is of the order of 1000 y.

Organic carbon is a minor component of both oceanic reservoirs, but it is important on long time scales because it is responsible for the formation of oceanic sediments from the skeletal remains of shell-forming aquatic biota, such as corrals.

The majority of oceanic carbon is inorganic: carbonic acid, $\sim 0.6\%$; bicarbonate ions, $\sim 90\%$; and carbonate ions, $\sim 9\%$. The carbonate ions can combine with calcium cations (Ca^{++}) to form calcium carbonate ($CaCO_3$), the form in which it can be removed from solution. The reservoir contents for both organic and inorganic carbon together are 6.7×10^2 PgC for the mixed layer and 3.7×10^4 PgC for the deep ocean.

Sedimentary reservoirs

The bulk of the earth's crust consists of old, igneous rocks that can be broken up by weathering; the weathered material is washed into the oceans to form sedimentary rocks, limestones, and sandstones. Since the planet first formed, approximately 8% of the planet is estimated to have been converted into sedimentary rocks. In the course of their formation the sedimentary rocks have been enriched in organic carbon from the biosphere. Igneous rocks average about 0.01% organic carbon, while the sediments average 3.6%; in the present epoch the weathering of sedimentary rocks cycles more organic carbon than does the weathering of igneous rocks.

The size of the continental sedimentary reservoir of organic carbon is 1.0×10^7 PgC. This is the source of the fossil fuels, coal, gas, and petroleum. Exploitable fossil fuels are estimated at 7×10^3 PgC, but the rapid use of this small component by humans gives it major importance for the carbon cycle (see equation 7.29), at least while it lasts. The current rate of consumption of fossil fuels is 6 PgC y^{-1} (Figure 7.3) and is rapidly increasing.

7.3.3 Cycles

Geological cycles

The terms *short*, *intermediate*, and *long*, as used in Table 7.1, are relative to the time scale for anthropogenic carbon emissions to fill the atmosphere ($\tau_i = \frac{G_i}{F_{ij}} = \frac{750 \text{ PgC}}{6 \text{ PgC y}^{-1}} = 125$ y). On the time scale of 125 y we are not concerned about the geological cycle, but a brief review is appropriate for completeness.

We consider net exchanges between the solid earth and the atmosphere. In this respect, the process involving most transport of carbon, the weathering of carbonate rocks, forms part of a closed cycle. The weathered carbonate is carried to the oceans together with calcium ions; subsequently, calcium carbonate is precipitated and reincorporated into the sediments. The sediments are later brought to the land by subduction and uplift. If it is not interrupted, this cycle is closed, and there is no net transfer of carbon dioxide to the atmosphere.

The weathering of silicate rocks also forms part of a closed cycle. Carbon dioxide from the atmosphere combines with silicon in the rocks and is washed into the oceans as bicarbonate and silicon ions. Silicon does not precipitate as carbonate so that when the silicon does form sediments the carbon dioxide is released to the oceans, and eventually to the atmosphere, on the time scale of oceanic transports.

This brings us to processes that do affect the atmosphere. The weathering and oxidation of rocks containing organic carbon does lead to a change (an increase) in atmospheric carbon dioxide. Reduced carbon, in whatever form, can combine with oxygen to form carbon dioxide, which enters the atmosphere. The rate has been estimated to lie between 10 and 100 TgC y^{-1}. The combustion of fossil fuels is one aspect of this process. The current rate of consumption of fossil fuels exceeds the natural process by a very large factor; indeed, the flux of industrial carbon outweighs that from all other natural geological processes. From an immediate point of view, this is the only process of any importance, but the geological steady state in the absence human intervention is, nevertheless, something that we need to understand.

The next cycle that we consider is perhaps the most interesting. The first step involves the weathering of calcium aluminum silicates (Figure 7.8), which can react with CO_2 as follows:

$$\underbrace{CaAl_2Si_2O_8}_{\text{anorthite}} + 3H_2O + 2CO_2 \rightarrow \underbrace{Al_2Si_2O_5(OH)_4}_{\text{kaolite}} + Ca^{++} + 2HCO_3^-. \quad (7.31)$$

The calcium and bicarbonate ions are washed into the oceans, where they are incorporated into sediments as calcium carbonate. These sediments are moved by subduction and uplift until they are again exposed to weathering. But the lost carbon is not recovered because, as we have seen, the weathering of limestone is a net-nul process for exchange with the atmosphere. The process suggests a net loss of both carbon dioxide and anorthite.

Finally there is volcanism. Although sedimentary carbonates, if undisturbed, will cycle without changing atmospheric carbon dioxide, the position will change if the carbonates come into contact with hot molten rocks in areas of volcanic activity. High temperatures can release carbon dioxide

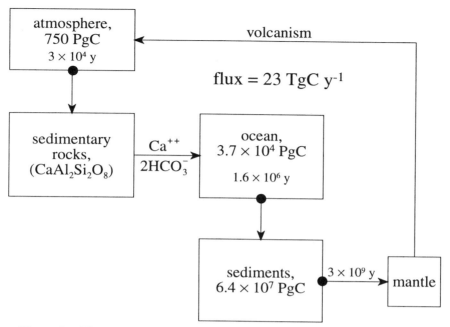

Figure 7.8 The calcium aluminum silicate cycle. The figure has been constructed on the basis of a flux through the cycle of 23 TgC y^{-1}. Time constants in years refer to the reservoir with the filled circle.

from carbonates, and volcanic gases contain 10% carbon dioxide, on average. It is estimated that 23 TgC y^{-1} enters the atmosphere by this route and compensates for the anorthite loss. The cycle is represented in this way in Figure 7.8.

Oceanic cycles

It is convenient to look at the intermediate and short oceanic cycles at the same time; they are compared with the intermediate and short biospheric cycles in Figure 7.9. Oceanic carbon is essentially inorganic, in contrast to biospheric carbon. The two groups communicate only through the atmosphere, but they can do so fairly rapidly.

Oceanic time constants have been calculated from measured concentrations of the radioactive isotope ^{14}C by the following method. The unstable isotope ^{14}C is formed naturally in the upper atmosphere by cosmic ray spallation and by nuclear explosions; it has a lifetime for natural, radioactive decay of $\tau_r = 5730$ y. It diffuses downwards in the form of ^{14}CO$_2$—rapidly through the atmosphere and more slowly through the oceanic mixed layer—

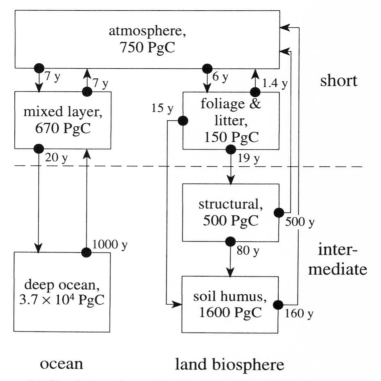

Figure 7.9 Biospheric and oceanic cycles. Compare with Table 7.1. Time constants apply to the reservoir with the filled circle.

and, for the most part, it decays in the deep ocean. If a steady state exists, the flux into the ocean must equal the decay rate in the ocean,

$$\frac{^{14}G_a}{^{14}\tau_{am}} - \frac{^{14}G_m}{^{14}\tau_{ma}} = \frac{^{14}G_m + ^{14}G_d}{\tau_r} \approx \frac{^{14}G_d}{\tau_r}. \tag{7.32}$$

The superscript 14 denotes the ^{14}C isotope, and the subscripts a, m, and d represent the atmosphere, the mixed layer, and the deep oceans, respectively; thus, τ_{am} is the time for the atmosphere to empty to the mixed layer, while τ_{ma} is the time for the reverse process, and the two are not the same. The Gs are carbon reservoirs.

The steady state for the stable $^{12}CO_2$ molecule has no decay term,

$$\frac{^{12}G_a}{^{12}\tau_{am}} = \frac{^{12}G_m}{^{12}\tau_{ma}}. \tag{7.33}$$

The transports depend upon molecular diffusion, which, for otherwise identical species, depends only upon the molecular mass. Laboratory measure-

ments indicate that
$$\frac{^{14}T_{ma}}{^{14}T_{am}} = 1.016\frac{^{12}T_{ma}}{^{12}T_{am}}. \tag{7.34}$$

The above equations yield

$$^{14}T_{am} \approx {}^{12}T_{am} = T_r \frac{^{14}G_a}{^{14}G_d}\left(1 - \frac{1}{1.016}\frac{^{14}G_m^{12}G_a}{^{12}G_m^{14}G_a}\right). \tag{7.35}$$

Equation (7.35) was obtained from the exchange between the atmosphere and the mixed layer. A similar equation may be obtained by assuming a steady state between the mixed layer and the deep ocean,

$$^{14}T_{md} \approx {}^{12}T_{md} = T_r \frac{^{12}G_m}{^{14}G_d}\left(\frac{^{14}G_m^{12}G_d}{^{12}G_m^{14}G_d} - 1\right). \tag{7.36}$$

Prior to nuclear testing it was possible to make measurements on the unperturbed state of the atmospheric and oceanic reservoirs. Assuming these to have been in a steady state, the following times were calculated:

$$T_{am} \approx 7.5 \text{ y},$$
$$T_{dm} \approx 1000 \text{ y}.$$

Biospheric cycles

The short and intermediate land, biospheric cycles are illustrated in Figure 7.9. Note that the atmosphere forms an essential link in these cycles. The short-term reservoirs of organic carbon can empty and fill 10 to 100 times faster than the intermediate reservoirs, with the consequence that exchange between the short-term reservoirs, that is, the atmosphere, the foliage, and the litter, is almost closed upon itself. This fast biospheric cycle is shown in Figure 7.10.

7.3.4 Global change

Access to the carbon in the crust and mantle is possible through weathering, tectonic processes, and the cycling of sedimentary rocks. Tectonic processes take about 10^8 to 10^9 years. The equilibration of the crust with the oceans, the atmosphere, and the biosphere on these times scales should give the "geological" background concentration of CO_2. According to some measurements this varies between 150 and 350 ppmv. This variation in background concentration should not be surprising because geological processes are themselves irregular, and with such a large dog and such a small tail large excursions in the concentrations of the smallest reservoirs are likely.

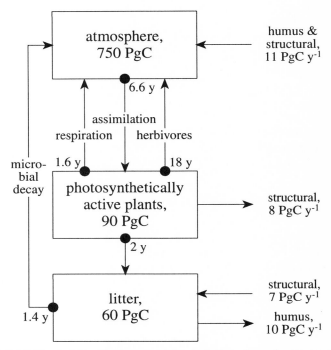

Figure 7.10 The fast biospheric cycle. The deviations from closure are shown by the fluxes on the right side of the figure. The times refer to the reservoir with the filled circle.

Some of the observed changes appear to correlate with the ice ages over the past 10^6 y. A statistical correlation does not establish cause and effect. Geological events may have caused the carbon dioxide variations, which, through the radiative greenhouse effect, may have led to the ice ages. The ice ages may have a different cause, for example, changes in solar radiation, and the resulting ice sheets may have changed surface weathering sufficiently to alter the carbon dioxide concentration. Note that the time for continental sediments to fill the atmosphere (Table 7.1) is less than the length of an ice age ($\sim 10^5$ y), so that the latter process is feasible.

If we turn now to the present epoch, Arrhenius, toward the end of the last century, first suggested that the burning of fossil fuels could lead to increases of carbon dioxide that could affect global temperatures. We now have several decades of data establishing the reality of the rise in carbon dioxide concentration. This is one of the few firm facts in the global change controversy.

We have noted that the observed rate of change in the atmospheric carbon dioxide concentration is approximately one half of that expected from the size of the industrial source. This does not cast doubt on the industrial origin of the observed increase, but it does raise the question of where the other half is stored. This question has obvious policy implications. Table 7.1 offers some useful data.

Time constants associated with the sediments and much of the deep ocean are too slow to be important when the effluent time scale is 125 y. At the other extreme, the short-term land biosphere is a small reservoir. If this were where the missing carbon dioxide is stored, it would soon be noticed. If all of the short-term biospheric reservoir were to enter the atmosphere, the amount of carbon dioxide would increase by only 20%.

The intermediate land biosphere, trees and humus, is a possibility for storing carbon. This reservoir is three times larger than the atmospheric reservoir, and the transfer time is similar to that for the transfer of fossil fuels to the atmosphere. Details need to be established, but it is plausible that the excess carbon dioxide has gone into this reservoir. If so, concern with the fate of the remaining forests is justified, as are the movements to increase forestation. If all of the world's trees were to burn, the atmospheric carbon dioxide could increase by a factor of almost 2. For all of the world's supply of fossil fuels to be taken up by forests and vegetation, they would have to increase in area by a factor of 10.

The upper ocean, including the mixed layer and a part of the deep ocean, provides a reservoir of similar size, with similar accessibility to the land biosphere. The missing carbon dioxide could be sequestered in the upper oceans. Schemes for increasing the rate of uptake of carbon dioxide into the oceans have been proposed, but they would be irresponsible in our present state of knowledge. It would be safer, but less popular, to reduce the consumption of fossil fuels, which would have additional benefits, for example, a decrease in acid rain.

7.4 Reading

Warneck's (1988) book, Chapter 1, is the best source for information about tropospheric chemistry.

7.5 Problems

Asterisks* indicate a higher degree of difficulty.

7.1 **Source of hydroxyl.** The tropospheric source of hydroxyl is thought to be photolysis, equation (7.5), followed by the reaction, (6.27), between $O(^1D)$ and H_2O. $O(^1D)$ can be thermalized in collisions with O_2 and N_2 molecules. The following data may be used at a temperature of

288 K in the planetary boundary layer:

$$[O_3] = 5 \times 10^{17} \text{ m}^{-3},$$
$$[H_2O] = 2.5 \times 10^{22} \text{ m}^{-3},$$
$$[N_2] = 1.99 \times 10^{25} \text{ m}^{-3},$$
$$[O_2] = 5.33 \times 10^{24} \text{ m}^{-3},$$
$$J_3 = 3 \times 10^{-4} \text{ s}^{-1}.$$

The Arrhenius coefficients for the reactions (see Appendix H, equation H.9) are given in the following table:

Reaction	k_0, m^3 s^{-1}	$\frac{E_a}{r_m}$, K
$O(^1D)+H_2O \to 2OH$	2.2×10^{-16}	0
$O(^1D)+N_2 \to O(^3P)+N_2$	1.8×10^{-17}	-107
$O(^1D)+O_2 \to O(^3P)+O_2$	3.2×10^{-17}	-67

(i) What fraction of $O(^1D)$ reacts to form hydroxyl?

(ii) What is the concentration of $O(^1D)$ in a steady state?

(iii) What is the production rate of hydroxyl (also when $O(^1D)$ is in a steady state)?

7.2 Lifetime of OH in the continental boundary layer. Hydroxyl is destroyed in bimolecular reactions with CH_4, CO, H_2, NO_2, HCHO, and other less important species. Rate coefficients for these reactions and concentrations of reacting species are listed in the following table:

Gas	Number density, m^{-3}	Rate coefficient, m^3 s^{-1}
CH_4	3.3×10^{19}	7.7×10^{-21}
CO	6.3×10^{18}	2.7×10^{-19}
H_2	1.5×10^{19}	7.5×10^{-21}
NO_2	5.0×10^{16}	1.1×10^{-17}
HCHO	5.0×10^{16}	1.2×10^{-17}

(i) What percentage of the total loss of hydroxyl is attributable to formaldehyde?

(ii) What is the lifetime of a hydroxyl molecule?

(iii) What steady-state concentration of OH is consistent with the result of Problem 7.1?

7.3 Measurement of OH. It is difficult to measure hydroxyl directly, but its concentration can be inferred from the growth rate of an anthropogenic gas with which hydroxyl reacts. CH_3CCl_3 and $CHCl_3$ are examples. If Q_X is the source of species X, both of these gases have sources that grow according to

$$Q_X = a_X \exp(b_X t).$$

Both species are removed by the reactions,

$$[X] + [OH] \rightarrow \text{products},$$

giving an inverse lifetime for X of

$$\tau_X^{-1} = k_X [OH].$$

(i) If we suppose that k_X and the total amount of OH in the entire atmosphere are constant, show that, after a sufficiently long time, the ratio of the observed amount of X to the amount emitted since $t = 0$ is,

$$R_X = \frac{b_X \tau_X}{b_X \tau_X + 1}.$$

(ii) Hence calculate [OH] from the observed data in the following table:

Trace gas, X	b_X, y^{-1}	R_X	k_X, m^3 s^{-1}
CH_3CCl_3	0.166	0.58	8.1×10^{-21}
$CHCl_3$	0.121	0.15	8.2×10^{-20}

7.4 Ca and Sr cycles. Calcium and strontium are both members of Group IIA of the Periodic Table and have many chemical similarities. Both are weathered from surface rocks and are washed into the oceans as anions accompanying carbonate and bicarbonate cations (see the cycle in Figure 7.8). Both elements are incorporated into the shells of marine organisms and subsequently sink to the bottom of the oceans to form oceanic sediments. The ratio Sr/Ca in shells is one-fifth the ratio in the surrounding sea water. Assuming that there is a steady state,

(i) How does the ratio Sr/Ca in river water compare with that for sea water?

(ii) What is the ratio of residence times for Sr and Ca in the oceans?

7.5 ^{14}C from nuclear explosions.* Before 1965, nuclear explosions injected large amounts of the radioactive isotope ^{14}C into the atmosphere. Before 1965 the level in the atmosphere rose and the level in the oceanic mixed layer adjusted to a new steady state in 2 or 3 y. Since 1965 the ^{14}C in the atmosphere has decayed but with a decay time longer than 10 y. All of these times are short compared with the natural decay time of ^{14}C (5730 y), so that natural decay does not play a part this discussion. Following the discussion (and notation) in §7.3.3, show that

(i)

$$\frac{d\Delta G_a}{dt} = \frac{\Delta G_m}{\tau_{ma}} - \frac{\Delta G_a}{\tau_{am}},$$

$$\frac{\Delta G_m}{dt} = \Delta G_a \left(\frac{1}{\tau_{am}} - \frac{1}{\tau_{dm}}\right) - \Delta G_m \left(\frac{1}{\tau_{ma}} + \frac{1}{\tau_{dm}} + \frac{1}{\tau_{md}}\right),$$

where ΔG_m and ΔG_a are *departures* of reservoirs of ^{14}C from their *final* values.

(ii) From the time constants given in Figure 7.9, calculate the two decay times that are associated with both ΔG_m and ΔG_a (the amplitudes can also be calculated, but this requires much more algebra).

7.6 **Deposition of sulfur.** The author discovered errors in the presentation of Figure 7.4 by comparing deposition rates (bottom line in Figure 7.4) with calculations based on reservoir sizes and decay times. Repeat these calculations to discover whether any discrepancies remain.

CHAPTER 8

CLOUDS AND PRECIPITATION

Water affects the atmosphere in many important ways. In this chapter we discuss the small amounts of condensed water that go to form clouds and mists. Most importantly, clouds and mists dominate the radiative properties of the atmosphere, when they are present. Clouds are an intermediate in the formation of rain.

The processes of formation and evolution of clouds involve both large-scale and small-scale properties of clouds and cloud systems. Important large-scale parameters include the air temperature at the cloud base, the cloud thickness, the magnitude of the updraft, and the lifetime of the cloud. These properties differ for the main classes of clouds. The two most important classes are cumulus and stratus (which includes cirrus). Cumulus clouds form on short-lived convective updrafts. They are deep but narrow, with updrafts of tens of meters per second. They last for an hour or less. Stratus clouds have weak updrafts, measured in tens of centimeters per second, and they form in large-scale, synoptic systems. They are typically 1 to 2 km thick, and extend over thousands of kilometers laterally. They may exist for hours or for days.

Important small-scale properties of clouds include the sizes of droplets, the presence or absence of ice crystals, and the availability of the foreign nuclei that are needed to initiate condensation. The range of sizes of these small-scale elements is large: The smallest condensation nuclei have radii $\sim 0.1 \mu m$, cloud droplets form with initial radii $\sim 10 \ \mu m$, the smallest drops that can form light precipitation have radii of $\sim 100 \ \mu m$, and raindrops are measured in millimeters. About 10^6 initial droplets must come together if one raindrop is to form.

In water clouds, precipitation involves four stages: initial condensation, two stages of growth leading to raindrops, and a multiplication process involving drop fragmentation. Initial condensation depends upon the number of active condensation nuclei and the amount of available condensate. In rising air, not all of the vapor condenses and a small supersaturation re-

mains.

The first growth stage involves condensation and the slow diffusion of supersaturated vapor through the boundary layer that surrounds the drop. Growth is inversely proportional to the drop radius, and it is difficult to obtain drops with radii larger than 20 µm by this process. The second growth stage involves differential fall speeds of drops with different sizes, leading to collisions and coalescence. For droplets with radii larger than 20 µm, coalescence takes over from condensation as the important growth mechanism.

The extent of growth by coalescence depends on the lifetime of the cloud or on the time that a droplet spends in a cloud before falling out as rain, whichever is smaller. The latter time depends on the strength of the updraft, because an updraft can hold a drop in suspension, and the depth of the cloud, because it sets the maximum length of a droplet trajectory. Deep cumulus clouds with strong updrafts can develop rain drops by coalescence. Once formed, large drops can break up, and the growth process repeats with the fragments, leading to a multiplication mechanism that may produce a shower.

Between $0°C$ and $-30°C$ clouds usually contain a mixture of ice crystals and supercooled water droplets. In such a disequilibrium mixture, the supersaturation of the vapor with respect to the ice phase can be very large, and condensation on ice crystals can be relatively rapid; drizzle droplets can form in partially frozen stratus clouds, even without coalescence. This is the Wegener-Bergeron precipitation process. However, growth in freezing clouds can also involve coalescence. Ice crystals falling through supercooled water droplets can form hailstones and dendritic crystals can interlock to form snowflakes. Both mechanisms can lead to fragmentation and to a multiplication process.

Given the sizes and phases of cloud particles, the bulk optical properties of a cloud may be calculated. Optical thickness for both solar and thermal radiation can be very large: The mean free path for radiation may be as short as 10 m. The two radiation streams differ principally in the single-scattering albedos of the droplets at solar and thermal wavelengths. Clouds have high surface albedos for solar radiation but are almost black for thermal radiation. These optical properties strongly influence climate.

8.1 Introduction

8.1.1 Cloud forms

To the experienced observer the taxonomy of clouds conveys an important level of understanding. The canonical classification elements—shape, height, and appearance—were not chosen to define the physics and dynamics of cloud formation, but they bear a close relationship to airflow, water content, and phase. A successful cloud physicist needs to be a dedicated observer.

Clouds are conventionally classified into 10 *cloud genera*, see Appendix I. *Stratus* and *cirrus* are *sheet clouds*, typically 1 to 2 km thick and hundreds of kilometers in lateral extent. They are associated either with widespread surface processes (condensation due to night-time cooling, or widespread convection) or with the organized lifting of large masses of air in depressions and in frontal systems. In strong contrast are the *cumulus* or *bunch clouds*, which result from concentrated convective activity. Cumulus clouds may rise 10 km or more above their base, and they have a similar lateral extent. The two major genera combine in the case of sheet clouds that contain smaller scale, convective structures, *stratocumulus* and *cirrocumulus*. *Orographic* clouds stand outside the main genera because they are fixed in space and are associated with geographical features.

In midlatitudes, sheet clouds may be classified into *low, middle,* and *high clouds*, with the height ranges 0 to 2 km, 2 to 7 km, and 5 to 13 km, respectively. High clouds usually contain at least some ice crystals, while low clouds usually consist only of water drops. Solid and liquid phases may be distinguished by their optical effects: Ice clouds tend to show color (sometimes), halos, and a fibrous texture; water clouds present a dense appearance with sharp outlines.

8.1.2 Microphysical properties of clouds

The *microphysical*[1] state of a cloud is determined by the dispersion of particle sizes, their phases (ice or liquid), the updraft in the cloud, and the turbulence. The last two dynamical factors will be discussed later, but here we shall introduce some data on droplet sizes and phases.

Particles occupy a spectrum of sizes. Low clouds consist of small water drops that fall slowly and give the impression of a quasi-stable suspension. Four liquid droplet spectra are illustrated in Figure 8.1. The two cumuli in the bottom panel have existed as entities for shorter times than have the other two clouds, and the droplet spectra are narrower. The broadening with time of droplet spectra is an important consideration for the onset of precipitation.

Medium and high clouds contain ice crystals, usually in a metastable mixture with supercooled water droplets. Not until the temperature falls below $-20°C$ (at 5 to 6 km in middle latitudes) are cloud particles predominantly ice. Ice particle concentrations vary from 10^{-3} m^{-3} to 10^{-2} m^{-3} in cirrus clouds and from $< 10^{-4}$ m^{-3} to 10^{-2} m^{-3} in cumulus anvils. Shapes

---
[1] Concerning the cloud particle structure. The term *macrophysical* is used to refer to the large-scale properties of clouds.

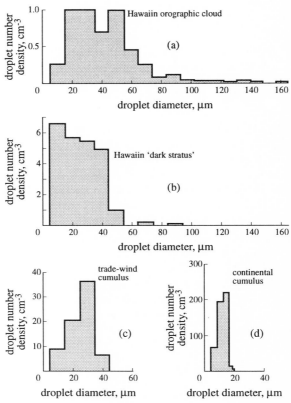

Figure 8.1 Droplet spectra. (a) Orographic cloud, $\rho_c = 0.40$ g m^{-3}; (b) dark stratus, $\rho_c = 0.50$ g m^{-3}; (c) trade wind cumulus, $\rho_c = 0.50$ g m^{-3}; (d) continental cumulus, $\rho_c = 0.35$ g m^{-3}. ρ_c is the density of condensed water, expressed in mass of water per unit volume of the atmosphere.

and sizes of ice crystals are highly variable. Needles and plates may be as large as 1000 by 40 μm or as small as 10 by 1 μm.

The transformation from a quasi-stable assembly of small drops to one that precipitates takes place in about 30 min. Rain drops typically have radii ~ 1000 μm. Because the radii of cumulus cloud droplets are ~ 10 μm, the obvious question is: How do 10^6 cloud droplets come together in such a short time to form a single raindrop? Similar questions can be framed about hail and snow. The microphysical aspects of such questions are partially understood and will be discussed in this chapter; limits to further progress in the understanding of precipitation are set more by the macrophysical or synoptic-scale characteristics of clouds and cloud systems.

CLOUDS AND PRECIPITATION

8.1.3 Condensation

Condensation occurs when air becomes sufficiently supersaturated in the presence of small *condensation nuclei*. Supersaturation can result from one of three processes: cooling due to adiabatic expansion in rising air, radiational cooling at night, and the mixing of two saturated air masses with different temperatures. The first is the most important. The rate of adiabatic cooling and the rate of condensation in an updraft are proportional to the speed of the updraft, which can be up to 50 m s^{-1} in convective clouds (cumulus) but is rarely more than 0.5 m s^{-1} in stratus. The relationship of the speed of the updraft to particle fall speeds, Appendix J, is crucial to the precipitation process and will be discussed in §8.4.

The amount of condensed water in a cloud is important for all cloud properties and behavior. An upper limit to observed water contents of clouds is provided by the *adiabatic water content* calculated, as the name implies, for an adiabatic updraft; it may be decreased by precipitation and by the lateral entrainment of drier air from the environment.

For adiabatic flow, the total water mass mixing ratio is conserved, $m = m_v + m_c =$ constant, where m_v is the vapor mixing ratio and m_c is the mixing ratio of condensate (see §2.2.1). Consequently,

$$\frac{dm}{dz} = \frac{dm_v}{dz} + \frac{dm_c}{dz} = 0 . \tag{8.1}$$

The moist entropy (see equation 2.27) is also conserved. In the limit $m \ll 1$,

$$\frac{ds}{dz} = \frac{c_{p,a}}{T_c} - \frac{r_a}{p}\frac{dp}{dz} + \frac{l}{T_c}\frac{dm_v}{dz} = 0 . \tag{8.2}$$

From equation (2.40) we may introduce the dry adiabatic lapse rate, and from (8.1),

$$\frac{dm_c}{dz} = \frac{c_{p,a}}{l}(\Gamma_{\text{dry}} - \Gamma) . \tag{8.3}$$

If conditions inside the cloud are hydrostatic in addition to being saturated, the lapse rate Γ is the saturated adiabatic lapse rate Γ_{sat},

$$\frac{dm_c}{dz} \approx \frac{c_{p,a}}{l}(\Gamma_{\text{dry}} - \Gamma_{\text{sat}}) . \tag{8.4}$$

At the cloud base $m_c = 0$, and (8.4) may be integrated upwards to give the adiabatic water content at any higher level, using data such as are given in Table 2.1. A calculation for a cloud base at 900 mb pressure is shown in Figure 8.2. The data are presented in the form of the mass of condensate per unit volume of air, $\rho_c = \rho m_c$, where ρ is the air density. For this cloud base, adiabatic liquid water contents are as high as 4 g m^{-3} in temperate latitudes and as high as 7 g m^{-3} in the tropics.

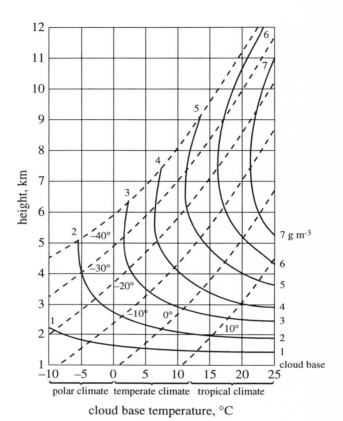

Figure 8.2 Adiabatic liquid water content. The base of the cloud is at 900 mb, about 1 km above the surface, corresponding to the bottom of the figure. The temperature of the cloud base is indicated on the horizontal axis. The broken lines show the temperature achieved at higher levels, based on climatological average conditions. The full lines show the adiabatic water content with numbers corresponding to densities in g m^{-3}. To use this diagram, select a cloud base temperature and travel along a vertical line starting from this temperature on the lower axis. At each level you can read off the height, the climatological mean temperature, and the adiabatic water content.

Returning briefly to other modes of condensation, radiational cooling is particularly important at the ground. During the daytime there may be a net radiative flux to the surface of several hundred W m^{-2}, which is usually reversed at night. The direct thermal effect of these diurnal changes in radiation is complex and is confined to the convective boundary layer (see Chapter 9), which averages a few hundred meters thick. In this layer diurnal temperature excursions can be large, and water evaporated during the day can condense at night in the form of dew or ground mist.

The third mode of condensation is illustrated by the misting of breath in cold weather and the mist flowing from an opened freezer compartment. If two saturated air parcels with different temperatures are mixed adiabatically at constant pressure, the mixture is always supersaturated, a result of the nonlinear dependence of saturated water-vapor pressure on the air temperature (see Appendix E).

8.2 Equilibrium between a droplet and its vapor

8.2.1 Kelvin and Raoult relations

Consider a system containing a water drop in equilibrium with its vapor, maintained at a constant pressure and temperature by external interactions; the derivation is similar with air present, but we omit this complication. The water droplet has a radius r, and a mass $M_l = M_w + M_s$, where M_w is the mass of water and M_s is the mass of *solute*, that is, the foreign matter dissolved in the drop. Water drops condense upon nuclei, and these nuclei may dissolve partially in the water.

The system is closed and the equilibrium condition for the entire system is, from equation (G.9),

$$dG = -SdT + Vdp = 0 . \tag{8.5}$$

Now look at the system in a different way, as the sum of three systems (or phases) consisting of liquid only (l), vapor only (v), and the surface separating them (s). We must bring in the surface phase because surface tension enables it to do work. The condition for equilibrium is

$$dG = dG_v + dG_l + dG_s = 0 . \tag{8.6}$$

Our discussion involves a number of implicit assumptions that we state but do not discuss: The air molecules, if any, remain in the gaseous phase and do not dissolve in the water, the solute molecules remain in the liquid phase, the volume of the surface system is negligible, and the total number of each molecule remains constant—the system is closed overall. However, each phase is not closed by itself because water molecules can move from one phase to the other. The differential chemical potentials for the individual phases must, therefore, include a transfer term, that is, the last term on the right of equation (G.8). When the three expressions are added, as in equation (8.6), the first two terms give the first two terms on the right of (8.5), which are zero. It follows that the sum of the three transfer terms must add to zero:

$$-\sigma dA + g_v dM_v + g_l dM_l = 0 , \tag{8.7}$$

where σ is the surface tension and the gs are partial chemical potentials.

Because water molecules are conserved, $dM_v = -dM_l$. From the geometry of a sphere, $dA = \frac{2dM_l}{r\rho_l}$, where ρ_l is the bulk density of the liquid drop and r is the radius of the drop. It follows that

$$g_v(r,T) - g_l(r,T) = \frac{2\sigma}{r\rho_l}. \tag{8.8}$$

If the surface were flat instead of curved, then $r = \infty$ and

$$g_v(\infty, T) = g_l(\infty, T). \tag{8.9}$$

This well-known result describes the condition for equilibrium between liquid and vapor for a flat liquid surface.

We see from (8.8) that some aspects of the chemical potentials are affected by the curvature of the drop. The chemical potential is a function of temperature and pressure, but the temperature is held constant for this discussion. It follows that the curvature affects the equilibrium vapor pressure. Equation (G.19) gives g_v in terms of the pressure. For the liquid phase, on the other hand, it is usually assumed that because the liquid is more or less incompressible, the chemical potential is independent of pressure. Taken together with equation (8.9), this implies

$$g_l(r,T) = g_l(\infty,T) = g_v(\infty,T). \tag{8.10}$$

Bringing together (8.8), (8.10), and (G.19), we find

$$p_v(r,T) = p_v(\infty,T) \exp\left(\frac{2\sigma}{rr_w T \rho_l}\right), \tag{8.11}$$

where r_w is the specific gas constant for water vapor. Because we are discussing a vapor in equilibrium with a liquid it follows that $p_v(\infty, T)$ is the same as $p_e(T)$, as used in equation (G.26) and tabulated in Table E.1. Equation (8.11) is known as the *Kelvin relation*.

The presence of a solute has repercussions. x_l, the molar mixing ratio of water in the liquid phase, is defined in equation (G.30). x_l must be less than unity and if the solution is dilute $(1 - x_l) \ll 1$. For dilute solutions we may use *Raoult's law*, equation (G.30),

$$p_e(x_l, T) = x_l p_e(x_l = 1, T) = x_l p_e(T). \tag{8.12}$$

Combining equations (8.11) and (8.12),

$$p_v(x_l, r, T) = p_e(T) x_l \exp\left(\frac{2\sigma}{rr_w T \rho_l}\right). \tag{8.13}$$

8.2.2 The Köhler relation

Expand (8.13) to first order, use the definition of x_l in equation (G.30), and approximate $x_l \approx 1$ to obtain the *Köhler relation*:

$$\frac{p_e(x_l, r)}{p_e} - 1 = S = \underbrace{\frac{2\sigma}{rr_w T \rho_l}}_{curvature\ effect} - \underbrace{\frac{3\nu M_s}{4\pi \rho_l r^3} \frac{\mathcal{M}_w}{\mathcal{M}_s}}_{solute\ effect} = \frac{A}{r} - \frac{B}{r^3}, \quad (8.14)$$

where $\mathcal{M}_w$ and $\mathcal{M}_s$ are the molar masses of water and of the dissolved solute. S is the *supersaturation*, usually expressed in percent, but only for $S > 0$. If the vapor pressure is less than the saturated vapor pressure, the related quantity, $\frac{p_v}{p_e}$, is called the *relative humidity* and is also usually expressed in percent.

The Köhler relation is plotted in Figure 8.3 for a range of masses of sodium chloride in solution. With the exception of pure water, each curve has a maximum at $r = r_c$, $S = S_c$, where

$$S_c = \left(\frac{4A^3}{27B}\right)^{\frac{1}{2}}, \quad (8.15)$$

$$r_c = \left(\frac{3B}{A}\right)^{\frac{1}{2}}. \quad (8.16)$$

For $r \ll r_c$, the solute effect dominates the vapor pressure (see equation 8.14); for $r \gg r_c$ the curvature effect is more important.

For $r < r_c$ the equilibrium expressed by (8.14) is stable, while for $r > r_c$ it is unstable. To see this, consider one of the four possibilities (loss or gain of liquid, $r > r_c$ or $r < r_c$), namely, a small loss of liquid from a drop for which $r < r_c$; the other three possibilities lead to the same conclusion. After the loss of water, the drop is reduced in size and finds itself to the left of and above the equilibrium curve in an environment with a higher water-vapor pressure than the equilibrium vapor pressure. Water molecules will diffuse toward the droplet, which will grow and return to its original size.

On this simple theory, droplets of pure water cannot grow directly from the vapor without infinite supersaturation (top curve in Figure 8.4). Statistical mechanical theory indicates that fluctuations can permit growth from the vapor phase (*homogeneous nucleation*) if the supersaturation exceeds 500% to 800%. The question is moot because atmospheric supersaturations rarely exceed 1% to 2%, for reasons that will emerge.

In practice, there are two reasons why water droplets may grow by condensation. First, condensation may take place on an insoluble nucleus of finite size. If the supersaturation is 2% and a nucleus has a radius of 0.6 μm, then, provided that water can wet the surface, the water molecules see the nucleus as water and a drop commences to grow.

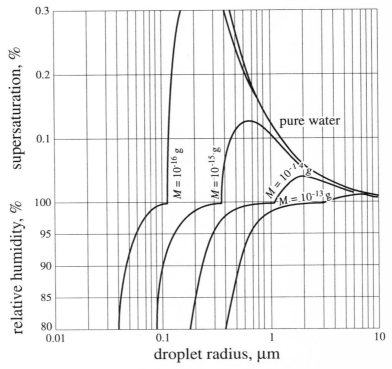

Figure 8.3 Equilibrium relative humidity or supersaturation. The nuclei are sodium chloride from sea water, and their masses are marked on the curves. Note the 100:1 change of scale at $S = 0$, or relative humidity 100%.

The second possibility is that soluble nuclei are present. Figure 8.3 suggests that growth is always possible on a soluble nucleus, no matter how small, but for drops to grow larger than r_c requires $S > S_c$. For a typical supersaturation of 0.47%, a salt crystal of mass 10^{-19} kg can lead to indefinite growth.

8.3 Condensation and freezing nuclei

8.3.1 Size distribution of aerosols

Insoluble *aerosols*[2] with radii smaller than 0.1 μm are relatively unimportant for cloud formation for the reasons given in the last section, except insofar as they can coagulate and form larger particles. These small nu-

---

[2]To be precise, all particles, solid or liquid, that are suspended in air are aerosols, but we restrict the term to particles that are neither water nor ice.

Table 8.1 Aerosol concentrations, m^{-3}

	Aitken	Large	Giant
Industrial city	10^{10}	10^{8}	10^{6}
Mid-ocean	$10^{8} - 10^{9}$	$10^{6} - 10^{7}$	10^{6}

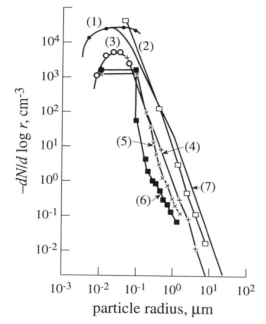

Figure 8.4 Size distribution of aerosol particles. (1) and (2) Frankfurt, (3) and (4) Zugspitze, (5) and (6) Crater Lake, (7) Seattle. The cumulative number density, N, is differentiated with respect to $\log r$, for reasons of convenience; see also equation (8.17).

clei are collectively referred to as *Aitken nuclei* after the inventor of the expansion-chamber technique used to detect them. More important for cloud physics are the *large nuclei* (0.1 μm $\leq r \leq$ 1.0 μm) and the *giant nuclei* ($r > 1.0$ μm). Most large and giant nuclei contain soluble components, particularly sodium chloride, that come from the oceans. Nuclei larger than 10 μm are rare because they fall rapidly and have only a short lifetime in the atmosphere.

Aerosol concentrations are extremely variable; orders of magnitude are shown in Table 8.1. Figure 8.4 shows typical aerosol size spectra at a variety

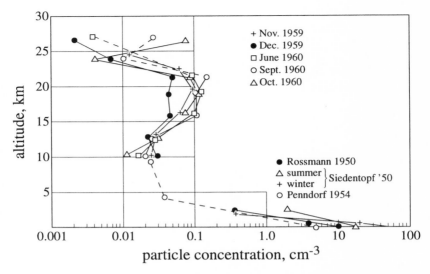

Figure 8.5 Vertical distribution of "large" aerosol particles, that is, all particles with radii greater than 0.1 μm.

of land locations. For $r > 0.1$ μm the slopes of the curves are similar,

$$\frac{dN(r)}{d\log r} \approx -cr^3 , \qquad (8.17)$$

where $N(r)$ is the *cumulative number density* of aerosol particles with radii larger than r and c is an empirical constant. Equation (8.17) is known as the *Junge distribution*. The distribution of aerosol number density with height varies with the size of aerosol particle. Giant nuclei are restricted to the convective boundary layer. Aitken nuclei are mixed more evenly to higher levels. Measurements of the vertical distribution of "large" particles are shown in Figure 8.5. Note the presence of the Junge layer, consisting of sulfate particles, near 20 km (see §7.2.3 and Figure 7.6).

8.3.2 Sources

A variety of physical and chemical processes are responsible for placing particles in the atmosphere. Volcanic debris is one obvious source. Soil and rock debris may be raised from the surface by winds, but with some difficulty. Sea salt enters the atmosphere by a complex process that involves first the breaking of a rising bubble and the subsequent development of a vertical jet that can reach 10 to 20 cm above the ocean surface. Drops as large as 2 mm in radius can be projected into the atmosphere. If they evaporate, they can leave behind a crystal of sodium chloride that is much smaller than the original droplet and may remain suspended in the atmosphere for long periods of time.

Forest fires produce organic vapors that are highly supersaturated and can nucleate homogeneously. Biological processes, particularly in hot, dry climates, can also produce supersaturated vapors that create a haze of condensed nuclei. Both processes create Aitken nuclei. Because of them, Aitken nuclei are much more frequent over land than over the oceans.

Gas-to-particle conversion is typified by the reactions discussed in §7.2.3, starting from COS and ending with sulfuric acid and ammonium sulfate aerosols. This reaction is a major source of atmospheric nuclei. Less well established is the possible importance of methanesulfonic aerosols resulting from the oxidation of dimethylsulfide, emitted from the oceans as a consequence of biological processes in the surface layers of the ocean (see §7.2.2). In a different category are the complex reactions involving chlorine nitrate and polar stratospheric clouds (see §6.7).

8.3.3 Cloud condensation nuclei

Only a fraction of the total number of dry aerosol particles serve as *cloud condensation nuclei*. From the standpoint of cloud formation, the important property of the atmospheric aerosol is its *activity spectrum*, the number density of particles that can act as condensation nuclei at specific supersaturations. Observers have reported relationships of the type

$$N(\text{ccn}) = cS^k , \tag{8.18}$$

where S is expressed in percent and $N(\text{ccn})$ is the total number of condensation nuclei of all sizes in m^{-3}. Typical values of the constants are $c = 5 \times 10^7$, $k = 0.4$ over the oceans, and $c = 4 \times 10^9$, $k = 0.9$ over the continents. These figures reflect the general fact that condensation nuclei are more frequent over land than over the oceans, with the result that maritime clouds have fewer but larger cloud droplets, with consequences for the probability of precipitation (see §8.7). For the continental case with $S = 1$, $N(\text{ccn}) = 4 \times 10^9$ m^{-3}, and if a liquid water content of 2×10^{-3} kg m^{-3} is shared equally between all nuclei, the average drop radius is ≈ 5 μm. This provides a starting point for a discussion of droplet growth.

8.3.4 Ice nuclei

The foregoing discussion of condensation nuclei concerned the formation of liquid drops. However, the ice phase plays an important part in precipitation processes and involves rarer types of nuclei. Ice crystals can form either by direct condensation or *sublimation* from the vapor phase or, alternately,

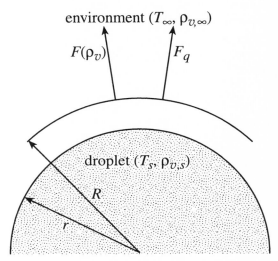

Figure 8.6 Growth by condensation.

by freezing water drops that have already condensed. The sublimation process requires nuclei, in the same way as for liquid condensation, the main difference being that *sublimation nuclei* are very rare, much rarer than condensation nuclei. The freezing of water droplets also requires the presence rare *freezing nuclei*. Pure water does not freeze spontaneously until about $-40°C$, but frozen cloud particles appear in clouds in increasing frequency from $-10°C$ to $-30°C$, and glaciation of clouds is usually complete before the temperature falls to $-40°C$. These figures indicate that rare nuclei are responsible for the formation *all* ice particles.

The distinction between sublimation and freezing nuclei is not crucial for our purposes, and they are collectively referred to as *ice nuclei*. Typical ice nucleus concentrations are 10 m^{-3} at $-10°C$ and 10^5 m^{-3} at $-30°C$, which may be compared with 10^6 to 10^7 m^{-3} for "large" condensation nuclei over the oceans.

8.4 Droplet growth in water clouds

8.4.1 Condensation

Isolated drop

A classic model that has been used for many decades to investigate droplet growth by condensation is shown in Figure 8.6. A droplet of radius, r, is placed in an environment for which the water-vapor density is $\rho_{v,\infty}$

CLOUDS AND PRECIPITATION

and the temperature is T_∞. The supersaturation of the environment is

$$S = \frac{\rho_{v,\infty}}{\rho_e(T_\infty)} - 1 \;. \tag{8.19}$$

From out discussion of the Köhler relation, illustrated in Figure 8.3, we require $S > 0$ for any droplet to grow indefinitely. However, the model that we shall develop is also valid for $S < 0$, or for relative humidity less than 100%, and the results may also be applied to the evaporation of a falling drop in the drier air below a cloud.

When a droplet is placed in the environment, molecular diffusion of water vapor and heat will take place through a boundary later surrounding the droplet. These fluxes simultaneously determine the rate of growth of the drop and also its surface temperature, T_s; the equilibrium vapor density at the surface, $\rho_{v,s}$, follows from the surface temperature. The relationship between the environmental densities and temperatures and the surface densities and temperatures will determine the directions of fluxes of water vapor, $F(\rho_v)$, and of heat, F_q.

The molecular diffusion equations that govern these fluxes are

$$F_q = -\kappa \frac{dT}{dR}, \tag{8.20}$$

$$F(\rho_v) = -D \frac{d\rho_v}{dR}, \tag{8.21}$$

where κ is the coefficient of heat conductivity for air, D is the diffusion coefficient for water vapor molecules through air, and R is the distance from the center of the drop.

Depending upon the sign of $F(\rho_v)$, water either condenses or evaporates at the surface of the drop, either releasing or absorbing latent heat. Provided that temperature differences are small, a steady state requires

$$F_q + lF(\rho_v) \approx 0 \;. \tag{8.22}$$

The solution to (8.20), (8.21), and (8.22) is

$$\frac{\rho_{v,\infty} - \rho_{v,s}}{T_s - T_\infty} = \frac{\kappa}{lD} \;. \tag{8.23}$$

Continuity of the flux of water molecules requires that

$$-4\pi R^2 D \frac{d\rho_v}{dR} = -\frac{dM}{dt} = -\rho_l 4\pi r^2 \frac{dr}{dt}, \tag{8.24}$$

where ρ_l is the bulk density of liquid water and M is the mass of the droplet. Equation (8.24) may be integrated with respect to R from r to ∞ to give

$$r \frac{dr}{dt} = \frac{D}{\rho_l}(\rho_{v,\infty} - \rho_{v,s}) \;. \tag{8.25}$$

At the surface of the drop, water vapor is saturated. From equation (8.14),

$$\rho_{v,s} = \rho_e(T_s)\left(1 + \frac{A}{r} - \frac{B}{r^3}\right). \tag{8.26}$$

The constants A and B are defined in equation (8.14).

Finally we adopt the approximate, integrated form of the Clausius-Clapeyron equation, (G.29), in the form,

$$\rho_e(T_\infty) - \rho_e(T_s) \approx \frac{l\rho_e(T_\infty)}{r_w}\frac{T_\infty - T_s}{T_\infty^2}. \tag{8.27}$$

r_w is the specific gas constant for water vapor. This approximation is valid when the temperature differences (and also water-vapor density differences) are very small.

With some manipulation of the above equations we may eliminate the temperature of the droplet and give the growth rate in terms of the environmental conditions and the droplet radius,

$$r\frac{dr}{dt} = \frac{S - \frac{A}{r} + \frac{B}{r^3}}{\rho_l\left\{\frac{r_w T_\infty}{D\rho_e(T_\infty)} + \frac{l^2}{\kappa r_w T_\infty^2}\right\}}. \tag{8.28}$$

Equation (8.28) represents either growth or decay, depending upon whether the numerator is positive or negative. In practice, the second and third terms in the numerator are usually small, so that the question principally concerns the sign of S. When the first term dominates the numerator the equation may be integrated to

$$r(t) = \sqrt{r^2(t=0) + ct}, \tag{8.29}$$

where c is half the right side of (8.28). Equation (8.29) suggests that after a certain time the memory of the initial drop size may be lost, and all drops tend to the same size. However, other growth processes take over before this can happen.

Tables 8.2 and 8.3 show data calculated on the basis of equation (8.28), the former for growth of a droplet in a slightly supersaturated atmosphere and the latter for evaporation of a droplet as it falls through unsaturated air beneath a cloud. The important lesson to be taken away from Table 8.2 is that it takes hours for particles to grow to radii in excess of 20 μm by diffusion alone. Clouds do not generally exist for hours. Table 8.3 shows that a droplet must exceed 0.1 mm (100 μm) in radius before it can survive 200 m below a cloud and stand a chance of reaching the ground. From these two observations we may conclude that droplet growth by diffusion from the vapor to the liquid phase is unlikely, by itself, to lead to significant amounts of rain.

Table 8.2 Growth of droplets by condensation

Nuclear mass, g	10^{-14}	10^{-13}	10^{-12}
Radius, μm	Time to grow from an initial radius of 0.75 μm, s		
1	2.4	0.15	0.013
2	130	7.0	0.61
5	1,000	320	62
10	2,700	1,800	870
20	8,500	7,400	5,900
30	17,500	16,000	14,500
50	44,500	43,500	41,500

The initial radii of the drops have been chosen to be 0.75 μm, and each contains the indicated mass of NaCl in solution. $T = 275$ K, $p = 9 \times 10^4$ Pa, $S = 0.05\%$.

Table 8.3 Distance fallen before evaporation

Initial radius	Distance fallen
1 μm	2 μm
3 μm	0.17 μm
10 μm	2.1 cm
30 μm	1.69 m
0.1 mm	208 m
0.15 mm	1.05 km

The atmosphere through which the drop falls has a temperature of 280 K and a relative humidity of 80% ($S = -0.2$).

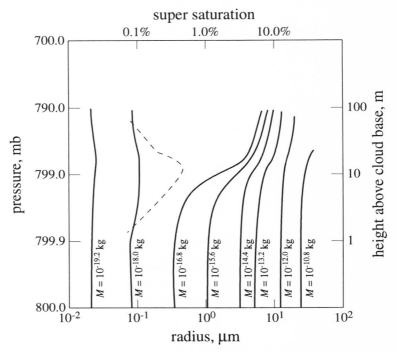

Figure 8.7 Growth of a mixed population of drops. An observed distribution of NaCl nuclei was chosen. The full lines show the droplet radii, starting with radii corresponding to the dry masses of the NaCl nuclei (marked against the full lines). The dashed line is the supersaturation. An updraft speed of 0.15 m s^{-1} was assumed.

Collective behavior

The state of supersaturation is not preordained. A cloud will tend towards a steady state in which the increase of liquid water in the form of droplets is balanced by the vapor condensed during the lifting process. Start with an assumed spectrum of soluble nuclei. Equation (8.28) may be used, given the supersaturation and the environmental temperature, to calculate the rate of growth of the droplets and the rate of supply of water that is required. The environmental temperature is given by the saturated adiabat if conditions at the cloud base are known. The rate of condensation in the updraft is given by equation (8.4) multiplied by the updraft velocity. From the balance of these two processes we may compute both the supersaturation and the droplet sizes.

An example of a numerical calculation is shown in Figure 8.7. The maximum supersaturation of 0.5% is achieved only 20 m above the cloud base. In this calculation the spectrum of nuclei was given, but if the number were determined by an activation process, depending on supersaturation, as for equation (8.18), all activation of cloud condensation nuclei would occur below this level of maximum supersaturation. The character of a cloud is, therefore, determined by the microphysical properties of the cloud very close to the cloud base and by the updraft.

On a number of occasions we have referred to the low supersaturations that actually occur in the atmosphere. The calculation shown in Figure 8.7 gives the essence of the reason why supersaturations are usually limited in size to a percent or less.

8.4.2 Coalescence

Water drops with different radii have different *terminal fall speeds*, w_t (see Appendix J). Spheres falling at different speeds can overtake one another; they make a geometric "collision" if their separation is less than the sum of the two radii, r_1 and r_2. Because, however, the smaller sphere tends to flow around the larger, the efficiency of the collision, $E(r_1, r_2)$, is less than unity. $E(r_1, r_2)$ is known from a mixture of experiment and theory (see Appendix J).

A spectrum or ensemble of drops of different sizes will coalesce or coagulate because of collisions, and the spectrum will change and evolve with time. This evolution is towards fewer, larger drops, and the ultimate question is how quickly raindrops may be assembled by this process.

When discussing coalescence it is common practice to picture the process in terms of larger drops, radii r_1, falling through and picking up smaller drops of radii r_2. Because both types of drop belong to the same ensemble, there is no real distinction between these two groups. Nevertheless, much use has been made of models based upon *continuous growth*, in which it is assumed that all r_1-drops are very much larger than all r_2-drops, so that the smaller drops can be treated as a continuum, to be swept up by the larger drops. We first give a rigorous expression for one aspect of the coagulation process and then approximate it to obtain the continuous-growth equation.

There is no loss of generality in assuming $r_1 > r_2$ as a matter of definition. In 1 s, the volume of space in which one r_1-drop can overtake an r_2-drop is $\pi(r_1 + r_2)^2 \{w_t(r_1) - w_t(r_2)\}$, where w_t is the terminal fall speed. If there is one r_2-drop per unit volume, the number of coalescences is

$$K(r_1, r_2) = \pi(r_1 + r_2)^2 E(r_1, r_2)\{w_t(r_1) - w_t(r_2)\}, \quad r_1 > r_2, \qquad (8.30)$$

where $K(r_1, r_2)$ is the *coagulation coefficient*. We now take the point of view that in a coalescence the volume of the larger drop is enhanced by the volume of the smaller drop, which disappears. If we integrate over all

r_2-drops,

$$\frac{d}{dt}\frac{4}{3}\pi r_1^3 = \int_0^{r_1} \frac{4}{3}\pi r_2^3 K(r_1,r_2) N(r_2)\, dr_2 ,\qquad (8.31)$$

where $N(r_2)dr_2$ is the number of drops with radii between r_2 and r_2+dr_2.

Equation (8.31) is not a complete statement of the coagulation process because it lacks a loss term for r_1-drops swept up by even larger drops, but it is the only term that matters for a continuous growth model. With $r_1 \gg r_2$ and $w_t(r_1) \gg w_t(r_2)$, (8.31) may be approximated to give the continuous-growth model,

$$\frac{dr_1}{dt} \approx \frac{w_t(r_1)}{3}\int_0^{r_1} \pi r_2^3 N(r_2) E(r_1,r_2)\, dr_2 = \frac{\overline{E}(r_1)\rho_c w_t(r_1)}{4\rho_l},\qquad (8.32)$$

where ρ_c is the *liquid water content* of the smaller drops,

$$\rho_c = \frac{4\pi}{3}\int_0^{r_1} r_2^3 \rho_l N(r_2)\, dr_2 ,\qquad (8.33)$$

ρ_l is the bulk density of water, and $\overline{E}(r_1)$ is a volume-weighted average of the collision efficiency.

Let us compare the continuous-growth model of coalescence with growth by condensation. For Stoke's flow (Appendix J), the terminal velocity is proportional to the square of the radius and, if r_2 is not small, $\overline{E} \approx 1$, so that

$$\frac{dr_1}{dt} \propto r_1^2 .\qquad (8.34)$$

The condensation mechanism, equation (8.28), for sufficiently large drops, gives

$$\frac{dr_1}{dt} \propto r_1^{-1} .\qquad (8.35)$$

As drops grow, sooner or later coalescence must overtake condensation, and experience suggests that the critical radius is ~ 20 µm. The rapidity of this transition is reinforced by the fact that $\overline{E}(r)$ increases rapidly as r increases through 20 µm. Because coalescence, once started, can proceed very rapidly, an important question in cloud physics becomes: What is the fastest route to a radius of ~ 20 µm? The answer to this question is strongly influenced by the presence or absence of the ice phase, and will be discussed further in §8.7.2.

It is usual to discuss coalescence in terms of height rather than time. To do so, multiply equations (8.31), (8.32), (8.34), and (8.35) by the droplet fall speed with respect to the surface, $w - w_t(r_1)$, where w is the updraft speed. The importance of the updraft is that droplets continue to rise until $w_t > w$. An updraft of 6.5 m s^{-1} can maintain droplets in a growing environment until they grow to a radius of 1 mm, provided, of course, that

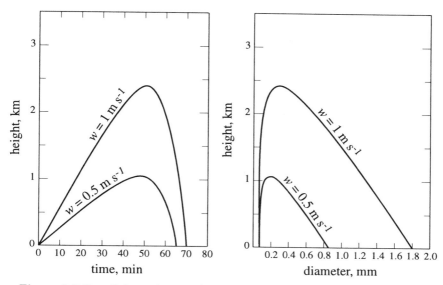

Figure 8.8 Growth by coalescence in cumulus clouds. (a) Drop trajectories, (b) drop diameters. The drop is assumed to have an initial radius of 20 μm and to fall through drops of 10 μm radius with a liquid water content of 10^{-3} kg m^{-3}. Updraft velocities are marked on the curves. They are typical of young cumulus clouds. Condensation is not included in this calculation.

the vertical extent of the cloud is sufficient to accommodate the trajectory of the drop (see Figures 8.8 and 8.9).

Calculations based upon continuous growth models are given in Figures 8.8 and 8.9. The former, for which the updrafts are typical of young cumulus, starts from a given drop size of 20 μm; condensation is not included in the calculation. Figure 8.9, for which the updraft is typical of stratus clouds, allows growth by condensation up to 20 μm radius and by coalescence above this radius.

Assuming that the cloud exists for long enough, the largest cloud drop that can be produced is the drop that falls from the bottom of the cloud. For cumulus parameters, millimeter-size drops are possible, although the time taken, more than 1 h in addition to the time to grow to 20 μm, is longer than the lifetime of a typical cumulus cloud. Parameters for a water stratus can lead to fine rain at best, with drop radii ∼ 100 μm.

The foregoing treatment of continuous growth dates from the 1950s and 1960s, and very much more sophisticated treatments of the coalescence process have been developed since then. Equation (8.31), with a loss term added, is an integral equation for which some solutions have been obtained. Generally speaking, for a full coagulation model the larger drops grow more rapidly than with the continuous growth model, and it is the largest drops

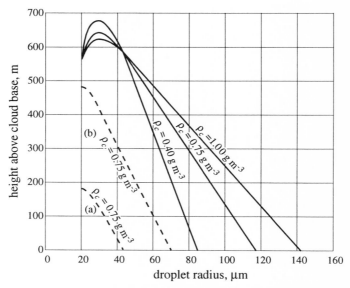

Figure 8.9 Growth by coalescence in stratus clouds. Assumed liquid-water contents are marked on the curves. Full lines: nuclear mass of NaCl = 10^{-16} kg, $T = 273$ K, $S = 0.05\%$, $w = 0.1$ m s^{-1}. Broken lines: nuclear mass of NaCl = 10^{-16} kg, $T = 273$ K, $w = 0.05$ m s^{-1}, (a) $S = 0.05\%$, (b) $S = 0.025\%$. Growth for droplets of radius less than 20 μm is calculated on the basis of condensation. For larger droplets growth is calculated by continuous coalescence.

that are most important for the development of precipitation (§8.7).

An additional factor that affects droplet growth is turbulence. Drops fall through a fluctuating rather than a steady vertical velocity, and the water content is distributed unevenly along the path. Both of these factors favor more rapid growth of larger drops.

The importance of large drops is greater than their small numbers would suggest. In §8.7 we shall consider drop-breaking mechanisms that can lead to a rapid increase in the number of large drops, once a few have formed. As far as precipitation mechanisms are concerned, it is the large-particle tail of the drop-size distribution that is important.

8.5 Optical properties

The discussion of §5.3.4 emphasized the great importance of clouds for climate. Clouds affect, to similar degrees, both thermal and solar radiation fluxes. Consider the extinction efficiencies for large spheres shown in Figure 3.12. A typical cloud droplet has a radius of 20 μm. For a wavelength of 10 μm and a refractive index ~ 1.33, the value of ρ is ~ 10. Going to

CLOUDS AND PRECIPITATION

shorter wavelengths means going to the right in the figure, and we may, therefore, anticipate similar extinction efficiencies for both thermal and solar radiation. Optical behavior in the two regions will differ because of different single-scattering albedos and different scattering asymmetry factors.

The bulk optical properties of ice and water show some similarities, particularly for wavelengths less than 10 μm. The real part of the refractive index does not vary greatly and lies between 1 and 1.6. More important are the large changes with wavelength of the complex part of the refractive index. Water and ice are both transparent in the visible spectrum; if we omit the ultraviolet spectrum, the complex part of the refractive index, n', is less than 10^{-4} for wavelengths less than 2 μm. At longer wavelengths the complex part rises steeply to as high as $n' = 1$, which is very large indeed. There are peaks corresponding roughly to the two fundamental vibration-rotation bands of water vapor and to the rotation band. To put the issue in terms of the formalism of Chapters 3 and 4, we may make use of a result from electromagnetic theory relating n' to the absorption coefficient,

$$k_\lambda = \frac{4\pi n'}{\lambda}. \qquad (8.36)$$

For a ray path along the diameter of a sphere, the optical path (equation 4.23) due to absorption alone is $4xn'$, where x is the size parameter $\frac{2\pi r}{\lambda}$ defined in equation (3.19). For $n' = 1$ only small particles, $x < 1$, will transmit any radiation of significance.

Mie theory calculations (§3.3.1) are appropriate for the droplets in water clouds. A calculation for cumulus clouds is shown in Figure 8.10. The data presented have been averaged over a typical drop-size spectrum. Calculations for water stratus do not differ greatly. An optical path is obtained by multiplying numbers in the top panel by the geometric path and the number density of cloud droplets, typically 10^8 to 10^9 m^{-3}; at a wavelength of 0.5 μm, an optical path of unity requires a geometric path of only 6 to 60 m. This is the mean free path of the radiation (see equation 3.6). It is very small compared with the thickness of a cloud deck. The data in Figure 8.10 show no substantial increase in the mean free path until the wavelength of the radiation exceeds 20 μm.

The data in Figure 8.10 have many implications. Very important is the qualitative difference between most of the solar spectrum and most of the thermal spectrum. For wavelengths greater than 2 μm the single-scattering albedo departs from unity and water droplets are strongly absorbing. For this reason it is common practice to approximate the thermal emission from a cloud by that of a black surface at the cloud-top temperature. A simple application of equation (4.63) shows that for this to be a good approximation requires that the single-scattering albedo be small, $a < 0.1$. This condition is only strictly satisfied for $\lambda > 70$ μm, but calculations with

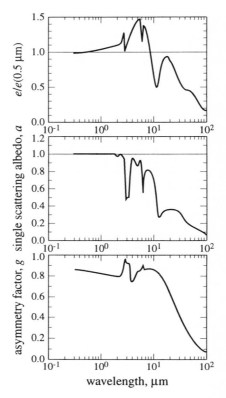

Figure 8.10 Optical properties of a cumulus cloud. All parameters have been averaged over a typical drop-size distribution. The top panel is the average extinction coefficient per droplet normalized to its value at $\lambda = 0.5$ μm, $e = 1.659 \times 10^{-10}$ m^2. The single scattering albedo is defined in equation (3.5) and the asymmetry factor in equation (4.61).

clouds as thermal black bodies are not seriously deficient when all other factors, including cloud variability, are considered.

The single-scattering albedo must be very close indeed to unity if a cloud is to have a high surface albedo for reflected solar radiation. Another application of equation (4.63) shows that, for isotropic scattering, a surface albedo $0.9 < A < 1.0$ requires a single-scattering albedo $0.998 < a < 1.000$. This condition is obeyed through most of the visible spectrum. But, to introduce a cautionary note into what is principally a theoretical subject, it should be noted that the data shown in Figure 8.10 lead to higher cloud albedos than are usually observed, particularly in the windows between absorption peaks. Much remains to be learned, even about water drops.

CLOUDS AND PRECIPITATION

8.6 Motions

The motions of clouds are of crucial importance to both cloud formation and to the initiation of precipitation. Cloud dynamics is beyond the scope of this book, but we include brief notes on three important classes of cloud to set the background for a discussion of precipitation in §8.7.

8.6.1 Cumulonimbus

Cumulonimbus clouds are huge convective systems that produce copious rain and usually hail and lightning. These clouds form at intense updrafts driven by the release of excess static energy in heated surface layers. Observations have established that organization of this concentrated flow requires convergence (and therefore upward motions) in the synoptic field in which the cumulonimbus is embedded. In other words, intense, local convection is triggered by large-scale events.

Updrafts have been measured up to 40 m s^{-1}. Cumulonimbus tops can rise to the tropopause and even penetrate a short distance into the stable stratosphere, forming a penetrative dome. The width of a cumulonimbus is typically twice its height.

It is usual to identify three stages in the life of a cumulonimbus. During the *cumulus stage* the cloud is dominated by individual updrafts, with downdrafts only in the upper sections and little precipitation. The *mature stage* (Figure 8.11) is the most active phase. Strong downdrafts accompany the updrafts and reach to the surface. These downdrafts can cause interactions between neighboring systems, and they may combine to form huge *mesoscale convective systems*, up to 100 km in diameter. At the cloud top the cloud is glaciated and the horizontal outflow is visible as a fibrous cirrus deck, known as the *anvil*. The most intense precipitation takes place in the mature phase.

In the *dissipating stage* the penetrative dome disappears. Downdrafts predominate, and the squall front (Figure 8.11) moves away from the cloud, taking the updraft with it. The three stages together last 45 to 50 minutes, of which the mature stage occupies 15 to 30 minutes. The time available for small cloud droplets to organize into large raindrops is short.

8.6.2 The cloud-topped boundary layer

In the absence of synoptic-scale convergence, the high static energy of surface layers heated by the sun will not organize into individual updrafts, as for cumulus clouds, but will give rise to a more uniform mixed layer or *convective boundary layer*, covering a large area. If the free atmosphere is stable, as is usually the case, this convective boundary layer will rise and

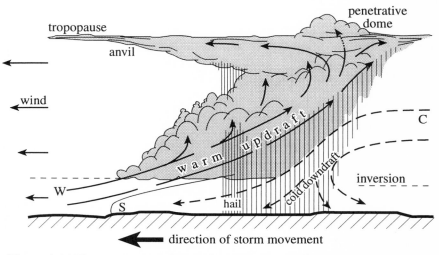

Figure 8.11 The mature stage of a cumulonimbus. Potentially unstable air is contained below an inversion. An advancing cold squall (S) forces the warm, unstable air (W) to rise, releasing energy for the cumulus convection. The cold air (C) flowing into the squall is further cooled by evaporating rain, giving rise to a strong downdraft. This is a two-dimensional projection of a three-dimensional flow. Neither the updraft nor the downdraft necessarily move with the ambient wind field.

be topped by a stable *inversion layer* (see Chapter 9 for further discussion). If the air is moist, the convective boundary layer will be capped by clouds, either stratus or stratocumulus. These clouds are important from the standpoint of the surface albedo, but they produce little, if any, rain. Ascent data for a cloud-topped boundary layer off the coast of California are shown in Figure 8.12.

8.6.3 Synoptic systems

A view from a satellite shows the extratropical regions to be more than half covered by vast sheets of mid- and high-level stratus clouds, associated with depressions and frontal systems. Mid-level layer clouds—altostratus and altocumulus—are about 3 km above the surface and consist mainly of water drops. High-level layer clouds—cirrus, cirrostratus and cirrocumulus—are about 10 km above the ground and are normally glaciated. The typical thickness of a stratus cloud is 1 to 2 km, and the updrafts are less than 0.5 m s^{-1}.

Large-scale systems can produce copious and persistent rain, but not with the violence of a cumulonimbus. A particularly well-documented case of frontal rain is shown in Figure 8.13. The frontal system is indicated

CLOUDS AND PRECIPITATION

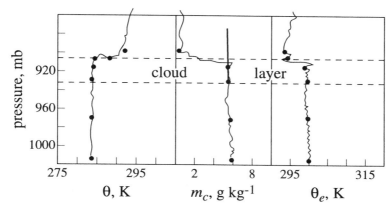

Figure 8.12 Potential temperatures and water-vapor mixing ratio in a cloud-topped boundary layer. The horizontal broken lines mark the top and bottom of the cloud layer. The radiometric temperature of the sea surface is 286.9 K. The equivalent potential temperature, θ_e, measures the moist entropy, §2.2.2. The negative gradient at the cloud top is unstable, (see §2.2.5), and the cloud top should be rising.

where it touches the surface. In the warm front, warm air rises over a colder wedge of air. In this particular example, the front sloped upwards by a vertical distance of ~ 3 km over a horizontal distance of ~ 250 km. Vertical velocities as high as 0.5 m s^{-1} were recorded.

8.7 Precipitation

8.7.1 Rain, snow, hail

Precipitation is conventionally divided into *drizzle, rain, showers, hail or graupel,* and *snow.* Drizzle consists of droplets of 100 to 150 μm radius and rainfall rates in drizzle are measured in millimeters per hour. It is the least activity that we call precipitation, and it is associated with the least active cloud systems. The term *rain* implies precipitation rates of centimeters per hour and drop sizes up to radii of 3 mm. Showers are short-lived, intense episodes of rain.

In the next section we shall find that it is important to know whether rain was ever in the ice phase. There is no internal evidence from the rain itself to tell us this, but the ice phase is obviously not involved when "warm" clouds (i.e., clouds whose temperatures never fall below freezing) precipitate. There are radar techniques that enable us to see whether precipitation has changed phase during descent. These observations indicate that melting commonly takes place in mid- and high-level clouds.

Heavy showers are sometimes associated with hail and graupel. The

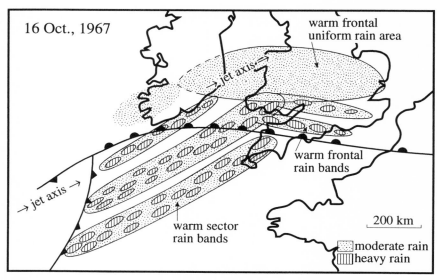

Figure 8.13 Rainfall in a wave depression. The system is steered by the jet stream. Moderate rain is about 2 mm h^{-1}; heavy rain is about 12 mm h^{-1}. The entire area of the figure was covered by stratus cloud.

size, mass and spherical form of hail and graupel show that they have grown by accreting supercooled water onto a falling ice particle.

Snowflakes indicate that growth has taken place only in the ice phase. The average snowflake consists of a number of separate dendritic crystals that aggregated during descent.

Microphysical models of droplet growth have been discussed in §8.4. To make this into a discussion of precipitation we must add three considerations: the significance of the ice phase, multiplication mechanisms for producing more than the occasional large drop, and macroscopic considerations connected with cloud motions. The third is the most complicated. The discussion of the last section shows how idealized are the simple models of §8.4. In fact, these models do no more than demonstrate the possibility of precipitation.

8.7.2 The Wegener-Bergeron process

Liquid water is readily supercooled into a metastable state. Ice and supercooled water can coexist but the two phases do not have the same vapor pressure, except at the triple point. At $-30°$C the vapor pressure over water is 34% higher than it is over ice; the relationship between degrees Celsius and excess vapor pressure is approximately linear. Thus, at any temperature below 0°C, an ice crystal surrounded by supercooled water drops experiences a large supersaturation, possibly 100 times larger than can develop with water drops alone (see Figure 8.7).

The importance of these facts for the growth of precipitation elements was first pointed out by Wegener in 1911; his ideas were later turned into a quantitative theory of precipitation by Bergeron, the *Wegener-Bergeron process*. Ice and supercooled water often coexist in clouds for almost any temperature between $-15°C$ and $-30°C$. The reason lies in the comparative rarity of freezing nuclei; above $-30°C$ there are simply not enough of them to freeze all of the droplets in a cloud.

The growth equation (8.28) is essentially the same for ice crystals except for a factor, of order unity, to allow for the shape of the crystal. For crystals that are large enough, equation (8.29) applies, with the constant c proportional to the supersaturation. For 100 times higher supersaturation, the radii in Table 8.3 should be multiplied by 10 and are consistent with the growth of drizzle size ice crystals within an hour. Partially glaciated mid-level stratus clouds can produce fine rain by condensation alone, without the need for coalescence. Condensation is, however, unlikely to take place without any subsequent coalescence, whether between pairs of ice crystals or between ice crystals and supercooled water.

The growth of ice crystals from the vapor phase is a complicated business with crystal habits that include plates, needles, and dendrites, depending upon the temperature and the supersaturation with respect to ice (Figure 8.14). An important point concerns the dendritic growth that develops between -12°C and -16°C. Dendrites fall extremely slowly, and their growth is favored compared to other forms because of the long time that they can spend in saturated regions of a cloud.

For intense precipitation there is not enough time available to grow every precipitation element from an initial nucleus. It is much more efficient to grow large elements from smaller elements that result from the fission of very large particles into a few (still large) components. *Multiplication mechanisms* are essential for any but the gentlest precipitation. For the Wegener-Bergeron process, multiplication results from the fission of snowflakes.

Large snowflakes, formed by linking together a number of dendrites, are extremely fragile and are readily fragment. Large fragments grow rapidly by condensation and by linking up with smaller flakes, and they can fragment in turn. The process repeats itself and many snowflakes develop from what was initially a small number.

The geometry of the cloud system matters for all forms of precipitation, including the Wegener-Bergeron process. Figure 8.11 illustrates how ice crystals from the anvil can seed the lower sections of a cumulonimbus cloud, with important consequences for the development of the cloud. This

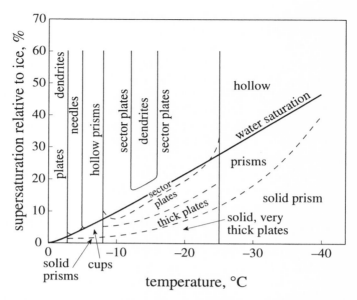

Figure 8.14 Ice-crystal habit as a function of temperature and supersaturation.

situation is a far cry from the simple models of particle growth that we considered in §8.4 and puts us on notice that simple models of atmospheric processes are rarely of more than qualitative significance.

8.7.3 Droplet coalescence

Scotch mist is a form of drizzle occurring on coastal hills. The uplift of air at the hill plays a role, and Figure 8.9 suggests that drizzle, but not heavier rain, may develop on a slight updraft. The time required for drizzle to develop is 2 h for the trajectories in Figure 8.9.

In tropical regions showers from "warm" cumulus clouds are common. Figure 8.8 illustrates how this may be possible. The clouds must be more than a few kilometers thick, the water content must be large, and the updraft must be on the order of 1 m s^{-1}. Such conditions often exist in cumulus clouds. We have pointed out that the calculations of §8.4 underestimate the rapidity of droplet growth. The important limitation may not be the coalescence mechanism but the time taken to produce initial droplets of 20 μm radius. This is where chance may enter. A rare, fortunate drop may start life with a particularly large, soluble nucleus, or alternately it may profit from a series of lucky chances, connected perhaps with complex cloud motions. Given a droplet multiplication mechanism, an occasional large drop may suffice to start a shower.

CLOUDS AND PRECIPITATION

The multiplication mechanism for water droplets involves the fracture of large drops, either spontaneously or when impacted by other drops. There are many different ways in which large drops can interact, with a wide variety of consequences. One set of wind-tunnel experiments on drops of radii 1.0 to 2.3 mm resulted in break up with the formation of up to 10 smaller drops, with an average number of 4. A 2 mm drop breaking into four fragments results in four 1 mm drops, which can themselves grow rapidly and, in turn, break up and multiply rapidly until the available liquid water in the cloud is exhausted.

8.8 Reading

A valuable elementary text is

Rogers, R.R., 1976, *A short course in cloud physics*. Oxford: Pergamon.

All aspects of clouds and rain are discussed in

Mason, B.J., 1971, *The physics of clouds*. Oxford: Clarendon Press.

A more advanced account of the microphysics is

Pruppacher, H.R., and Klett, J.D., 1978, *Microphysics of clouds and precipitation*. Dordrecht: Reidel.

Cloud dynamics are discussed in detail by

Cotton, W.R., and Anthes, R.A., 1989, *Storm and cloud dynamics*. New York: Reidel.

A view from the standpoint of an observer of clouds:

Ludlam, F.H., 1980, *Clouds and storms*. University Park, PA: The Pennsylvania State University Press.

8.9 Problems

Asterisks* and double asterisks** indicate higher degrees of difficulty.

8.1 **Weather radar.*** During World War II it was discovered that detectable radar scattering could be obtained from precipitation. In subsequent years, more powerful radars have detected scattering from non-precipitating cloud particles, and the measurement of Doppler shifts has permitted simultaneous measurement of velocities. Weather radar is now a powerful research tool.

The geometry of a weather radar is shown in Figure 8.15. The antenna sends out a pulse of length, l, and power, $\mathcal{P}_0$, into a solid angle, Ω; the radiation is scattered by particles in the illuminated volume and returns power, $\mathcal{P}_s$, into a solid angle, A/R^2. The illuminated volume may be assumed to be optically thin.

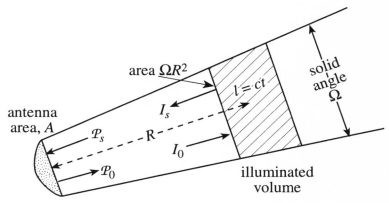

Figure 8.15 Geometry of a weather radar.

A = antenna area;
R = range;
I_0, I_s = incident, scattered radiances;
$\mathcal{P}_0, \mathcal{P}_s$ = radiated, received powers;
A/R^2 = solid angle filled by antenna;
Ω = solid angle of antenna;
l = pulse length.

From equations (4.2), (4.15), and (4.26) derive the weather radar equation,
$$\frac{\mathcal{P}_s}{\mathcal{P}_0} = \frac{I_s}{I_0} = \frac{Al}{4\pi R^2}\Sigma,$$
where
$$\Sigma = \int_0^\infty n(r)s(r)P(\pi,r)\,dr,$$
r is the droplet radius, $n(r)\,dr$ is the number of droplets between r and $r + dr$, $s(r)$ is the scattering coefficient, and $P(\pi, r)$ is the back-scatter phase function.

8.2 Radar reflectivity.

(i) Commonly used radar wavelengths are 3 and 10 cm. Calculate the size parameter, equation (3.19), for 10-μm radius cloud drops and 1-mm radius rain drops. Give an approximate expression for Σ based on the following relationship for the polarizability of a dielectric sphere in terms of the bulk refractive index of water, $\tilde{m}$:
$$p = \left(\frac{\tilde{m}^2 - 1}{\tilde{m}^2 + 2}\right) r^3.$$

Show that your result is consistent with that in Table 3.2.

(ii) If there is a bimodal distribution of 1 mm rain drops and 10 μm cloud drops, with 100 times more liquid in the latter than in the former, what is the ratio of the received powers from the two sources?

(iii) Express the rainfall in terms of the droplet spectrum assuming Stoke's fall. The radar function, Σ, is sometimes used as a measure of the rainfall. Is this justified?

8.3 The melting band. In two-dimensional radar reflectivity presentations (the Range Height Indicator or RHI) there often appears a layer of enhanced reflectivity corresponding to the $0°C$ level, at which snowflakes melt and turn to rain. There are two aspects of this *bright band* or *melting band* that need to be explained: Why does the reflectivity increase entering the top of the band; and why does it decrease leaving the bottom? Suggest answers for these questions.

When a snow crystal begins to melt it first appears to the radar to be water in the same shape as the ice crystal. Radar reflectivities are dominated by the real parts of the refractive indices of water and ice. These are 8.99 and 1.78, respectively, at $0°C$ and 10 cm wavelength.

8.4 Cloud height determination.** There are many techniques for estimating cloud properties from satellite observations. The CO_2-*slicing* method uses measurements in two contiguous fields of view (primed and unprimed radiances) and at two close thermal frequencies in the 667 cm^{-1} band (subscripts 1, and 2) to determine the cloud height. It is assumed that the two fields of view differ only in the fractional cloud cover (the clouds are generally smaller than the field of view, and they may be partially transparent), that the clouds do not scatter thermal radiation, that they are physically thin, that the relation between p and T is the same for clear and cloudy regions, and that the optical properties of the cloud are the same for the two frequencies, while the absorption of carbon dioxide differs.

Derive the fundamental relation for the CO_2-slicing method,

$$\frac{I_1 - I_1'}{I_2 - I_2'} = \frac{\int_{p_{ct}}^{p_s} \mathcal{T}_1(p) \frac{dB_1}{d\ln p} d\ln p}{\int_{p_{ct}}^{p_s} \mathcal{T}_2(p) \frac{dB_2}{d\ln p} d\ln p}.$$

p_s is the surface pressure and p_{ct} is the pressure at the cloud tops. $\mathcal{T}_{1,2}(p)$ are transmissions from space to the pressure level, p.

8.5 **The clouds of Venus.*** Venus is covered by a dense cloud layer, the main part of which lies between 47 and 70 km above the surface. Throughout the clouds the temperature decreases with height at about the same rate as in the terrestrial troposphere. In situ and earth-based observations suggest that we are dealing with three classes of aerosol. Mode I consists of submicron particles; Mode II consists of a narrow size spectrum with a peak near 1 μm radius. Mode III particles have radii between 4 and 15 μm.

Mode II probably consists of drops of very strong sulfuric acid, formed perhaps by photolysis of SO_2 following the reactions in equations (7.16), (7.17), and (7.18), taking place above the clouds. The size spectrum is amazingly narrow. The best measurements suggest $r = 1.05 \pm 0.1$ μm.

(i) Why do you think the cloud base is at 47 km?

(ii) Can you suggest why the Mode II size distribution is so narrow? What, if anything, does this suggest about the Mode III particles?

(iii) In the main cloud deck the rate of change with distance of optical depth is $\frac{d\tau}{dl} = 1 \times 10^{-3}$ m^{-1}. If this opacity is caused by Mode II particles, what approximately is their number density?

(iv) A mode III particle has a radius of 10 μm. What is the average time that it will take to fall out of the cloud? What is the fall time for a Mode II particle? (Assume that the atmosphere of Venus has the same viscosity as air. To a first approximation viscosity does not depend upon pressure.)

(v) A Mode III drop falls through Mode II drops and collects them up by a continuous process. If the collision efficiency $\overline{E}$ is constant during the process, find an expression for the increase in size of a Mode III drop after it has fallen through an optical path τ.

CHAPTER 9

THE PLANETARY BOUNDARY LAYER

The planetary boundary layer is that region of the atmosphere that is directly influenced by the presence of the earth's surface. A useful level of knowledge about the boundary layer has been built up based on empirical data, the general nature of the equations of fluid motion, and dimensional arguments.

Very close to the earth's surface the wind velocity is reduced to zero by the drag of surface elements. This takes place in the roughness layer, the depth of which is comparable to the size of the surface roughness elements. The flow above the roughness layer contains small-scale, time-dependent motions, or eddies. Velocities, temperatures, and other state variables may be expressed formally as the sum of mean variables and eddy variables. Dynamical meteorology cannot follow the eddy motions and works with equations that relate the mean variables. But the eddies cannot be ignored; their effect must be included in the form of eddy fluxes of momentum, entropy, etc. These fluxes (or their divergences) must be added to the equations of motion for the mean variables. Eddy fluxes can be large and important, and are usually treated in terms of a diffusion equation with large eddy diffusion coefficients.

Under the influence of eddy stresses, the free atmosphere winds slow down and turn as the surface is approached. This takes place in approximately the lowest kilometer of the atmosphere (in middle latitudes), in the Ekman layer. In the lowest 100 m or so stresses are approximately constant, and it is for this constant-stress layer that theoretical contributions have been most effective.

Eddies may be generated by shear forces or by convection (buoyancy). We discuss two general classes of theory: local dissipation, in which there is no transport of eddy energy, and generation is in the same place as dissipation; and nonlocal dissipation, for which eddy energy is generated in one part of the boundary layer and dissipated in another. Nonlocal dissipation is represented by a treatment of the diurnal rise of a surface inversion in the convective layer over a surface heated by solar radiation. For this treatment the constant-stress layer is treated as a single entity.

Other sections of this chapter deals with local dissipation, under both adiabatic (or neutral) and nonadiabatic conditions. For neutral flows eddies are generated by shear forces. An important foundation for the theory of the neutral boundary layer is provided by the classic wind-tunnel experiments of Nikuradse, which gave a logarithmic profile of velocity. Nonadiabatic flows involve either generation or destruction either by shear or by convection. Monin-Obukhov similarity theory deals with this problem in the constant-stress layer. The theory expresses the wind profile in terms of the ratio of the height to the Monin-Obukhov length, a quantity formed from the measurable entropy and momentum fluxes. It agrees remarkably well with observations made over a wide range of atmospheric conditions.

That boundary-layer theories agree well with observations reflects, in part, the empirical evidence used in the construction of the theories, but it also demonstrates that we have a working level of understanding of the physical processes involved in the planetary boundary layer.

9.1 Concepts

9.1.1 Classification of regions

The *planetary boundary layer* is that region of the atmosphere in which motions are directly influenced by the presence of the earth's surface. According to our simple model of §5.1.2, the entire troposphere is affected by the solar radiation absorbed at the surface and is, in this sense, a boundary layer, but the term is usually employed in a more restricted context (see Table 9.1).

We may quickly dispose of the first and last entries in Table 9.1. Between the tropopause and approximately 1 km above the ground (in middle latitudes), the stress acting on the mean wind is relatively small and the motions are close to geostrophic, or so we may assume (see equations F.14 and F.15). Only lower lying layers are said to belong to the planetary boundary layer.

If the only source of stress were molecular viscosity, this geostrophic regime would continue down to a *viscous sublayer*, whose thickness may be shown to be

$$\delta \sim \frac{5\nu}{u_{*,0}}. \tag{9.1}$$

ν is the kinematic viscosity, and $u_{*,0}$ is the surface value of the *friction velocity* u_*, defined by

$$u_* = \sqrt{\frac{\tau}{\rho}}, \tag{9.2}$$

Table 9.1 The atmospheric boundary layer

Region	Equations	Comment	Vertical scale
Free atmosphere	$\overline{u}_g \sim \frac{1}{2\Omega \sin \phi} \frac{\partial \overline{p}}{\partial x}$	Geostrophic	$H \sim 8$ km
Ekman layer	$\frac{\partial \overline{\tau}_{ij}^e}{\partial x_j}$ significant	Wind turns	~ 1 km
Diurnal mixed layer	Layer average	Time dependent	~ 1 km
Surface layer	$\overline{\tau}_{ij}^e = \tau_0$	Shear and buoyancy	~ 100 m
Neutral layer	Same	Only shear	$z < |L| \sim 30$ m
Roughness	Form drag	Rough flow	Obstacle height
Viscous layer	$\tau_0 = \overline{\rho}\nu \frac{\partial \overline{u}}{\partial z}$	Smooth flow	$\frac{5\nu}{u_{*,0}} \sim 0.5$ mm

Orders of magnitude

$\overline{\theta} \sim 300$ K $\qquad 2\Omega \sin \phi \sim 10^{-4}$ s^{-1}

$\overline{u}_g \sim 10$ m s^{-1} $\qquad \nu \sim 10^{-5}$ m^2 s^{-1}

$u_{*,0} \sim 0.3$ m s^{-1} $\qquad \overline{\rho} \sim 1$ kg m^{-3}

$\tau_0 \sim 0.1$ kg m^{-1} s^{-2} $\qquad c_p \sim 10^3$ J kg K^{-1}

$\frac{g}{\theta} \sim 3 \times 10^{-2}$ m s^{-2} K^{-1} $\qquad \frac{F_0(h)}{\overline{\rho} c_p} \sim 0.2$ K m s^{-1}

where τ is the internal stress (see Appendix F). At the surface this stress is τ_0, the *surface stress*. For the data in Table 9.1, $\delta \sim 0.5$ mm, a figure that is clearly irrelevant to the rough surface of earth. Drag at a rough surface is caused by *form drag* at surface obstacles in a *roughness layer*, and is called *aerodynamically rough flow*. The depth of this roughness layer is set by the heights of the roughness elements (grass, stones, etc.) on the surface.

Between the lowest layers, in which molecular effects can be felt, and the geostrophic region, in which stress can be ignored, lies the planetary boundary layer. Table 9.1 suggests that we need to consider four different, but not necessarily independent, regimes within the planetary boundary layer, each of which will be described in subsequent sections.

9.1.2 The logarithmic profile

Early wind-tunnel measurements by Nikuradse for aerodynamically rough flow established that the mean wind speed, $\overline{u}$, that is, the wind averaged over all fluctuations, varies with distance from the wall of the wind tunnel, z, according to

$$\frac{\partial \overline{u}}{\partial z} = \frac{u_*}{kz}. \qquad (9.3)$$

k is *von Karman's constant* ≈ 0.41. For a rough surface, the zero for z is indeterminate. It is usual to determine a *zero-plane displacement* from the surface from which to measure the height by means of an empirical fit to the data. This displacement is typically $2h_c/3$, where h_c is the height of the roughness elements that disturb the boundary-layer flow.

Close to the ground lies a region in which u_* is essentially constant and equal to its value at the surface, $u_{*,0}$. This region is known as the *surface* or *constant-stress layer*. If conditions throughout the surface layer correspond to those in the wind tunnel, (9.3) may be integrated to yield a *logarithmic profile* of wind with height,

$$\frac{\overline{u}}{u_{*,0}} = \frac{1}{k} \ln \frac{z}{z_0}. \qquad (9.4)$$

The constant of integration, z_0, is called the *roughness length* and is the height, measured from the zero-plane, at which $\overline{u}$ extrapolates to zero. As is done for the zero-plane displacement, z_0 may be derived empirically by fitting (9.4) to observed data. z_0 is very variable but is typically $\sim 0.1\,h_c$. Some examples from the literature are: soil, $z_0 \approx 1$ mm to 1 cm; thick grass, $z_0 \approx 2$ cm; pine forest, $z_0 \approx 0.3$ to 0.9 m.

In Nikuradse's experiments the boundary layer transmits momentum (a stress is a negative momentum flux, see Appendix F) but not heat. In this event, the flow is adiabatic and is called a *neutral flow*. Equation (9.3) lies at the basis of all treatments of neutral boundary layers in the atmosphere. It may be objected that, with two adjustable constants, a good fit to observations is not a discriminating test of the physics of the boundary-layer flow, and that is so. However, we are faced with a need to treat the boundary layer, combined with the fact that decades of research have not led to fundamental theories valid for rough terrain. For the planetary boundary layer we must be content with a mixture of arguments based on empirical knowledge, the nature of the equations of fluid motion, and dimensional arguments. Examples of each will follow.

9.1.3 Eddy fluxes

In the foregoing discussion we referred to stresses and to the friction velocity in a region of the flow where molecular viscosity is unimportant. At first

THE PLANETARY BOUNDARY LAYER

sight this appears to be inconsistent, but the rationale lies in the use of a mean flow, an idea that we introduced without discussion in equation (9.3). We must now examine that assumption with more care.

At all times and at all atmospheric levels there are time-dependent motions, often of smaller size than features in the mean flow that we wish to consider. These small-scale, time-dependent flows are referred to as *eddies*, regardless of their form: They may be quasi-random (turbulent) or they may be regular waves, but their effect is similar. For practical reasons, atmospheric dynamics cannot work with small-scale motions and therefore uses averaged equations. This act of averaging generates terms in the equations of motion that behave like stresses on the mean flow (*eddy stresses*).

Mathematically, eddy motions may be distinguished from the mean motions in the following way. We write

$$u = \overline{u} + u' . \tag{9.5}$$

$\overline{u}$ is the mean wind that we wish to retain, while u' is the eddy component. To distinguish between them requires a definable separation of scales of length or of time. In fact, Fourier decompositions of space or time series of atmospheric variables exhibit continuous spectra and do not show simple separations.

This is a difficult topic. In principle, a similar question arises with the distinction between fluid motions and molecular motions, but the scale separation is obvious in this case and leads to a practical distinction between internal energy, on the one hand, and fluid kinetic energy, on the other hand, which we tend to take for granted. Similarly, eddy motions may be treated as distinct from mean motions, and analogies may be drawn to molecular motions.

We shall assume that a scale separation exists between mean flows and eddies and that it defines the averaging process. The important result, which we shall repeatedly use, is that the average values of eddy components are zero,

$$\overline{u'} \equiv 0 . \tag{9.6}$$

Let $\chi = \overline{\chi} + \chi'$ be an extensive property of the atmosphere. From equation (F.21), a flux component may be written in the form

$$\begin{aligned}\frac{F(\chi)}{\rho} &= (\overline{u} + u')(\overline{\chi} + \chi'), \\ &= \overline{u}\,\overline{\chi} + u'\overline{\chi} + \overline{u}\chi' + u'\chi'. \end{aligned} \tag{9.7}$$

Apply the averaging operator to (9.7), and use (9.6),

$$\overline{\frac{F(\chi)}{\rho}} = \overline{u}\,\overline{\chi} + \overline{u'\chi'} . \tag{9.8}$$

If we compute a flux using only averaged variables (the first term on the right of 9.8), we are obliged to add an *eddy flux*, $F^e(\chi)$, corresponding to $\overline{u'\chi'}$ to complete the calculation. For example, because stress is a negative momentum flux, we must, if we wish to calculate with averaged variables, add an eddy stress, τ^e_{ij}, to the molecular stress tensor (which we now call τ^m_{ij}) that was used in equation (F.5). From (9.8) with $\chi' = u'$,

$$\frac{\tau_{ij}}{\rho} = \frac{\tau^m_{ij}}{\rho} + \frac{\tau^e_{ij}}{\rho} = \frac{\tau^m_{ij}}{\rho} - \overline{u'_i u'_j}. \tag{9.9}$$

$\overline{u'_i u'_j}$ is the *Reynolds' stress tensor*. The diagonal terms of the tensor are twice the average values of the *eddy energy* per unit mass,

$$k^e = \frac{1}{2} \overline{u'_i u'_i}. \tag{9.10}$$

The nondiagonal components are the eddy stresses,

$$\frac{\tau^e_{ij}}{\rho} = -\overline{u'_i u'_j}. \tag{9.11}$$

They are nonzero if orthogonal motions are partially correlated, which is commonly the case.

The magnitude of the eddy stress relative to the molecular stress is an important issue. In Table 9.1 we noted a typical boundary-layer stress of 0.1 kg m^{-1} s^{-2}. We may calculate the molecular stress averaged over the whole Ekman layer by multiplying the mean velocity gradient, 10^{-2} s^{-1}, by the kinematic viscosity and the air density: This yields $\tau^m \sim 10^{-7}$ kg m^{-1} s^{-2}. Under these circumstances, molecular effects can be neglected.

In the viscous sublayer the stress must be carried by molecular diffusion alone. In this thin layer, we may show from equations (9.1) and (F.6) that the velocity increases across the layer by $5u_{*,0}$, or 1.5 m s^{-1} for the data in Table 9.1.

When the wind has risen to ~ 1 m s^{-1}, shear instabilities are almost inevitable. They give rise to turbulent eddies that grow in size as the distance from the surface increases and are soon capable of carrying the entire stress. The reader may compare this discussion to that of surface heat and entropy fluxes in §2.3.6.

9.1.4 Boussinesq approximation

The following approximations are widely adopted, and we shall make use of them. They appear to be reasonable at face value, but detailed justification is not simple and the reader is referred to the books listed at the end of the chapter for details. The first two approximations are associated

THE PLANETARY BOUNDARY LAYER   259

with the name of *Boussinesq*; the third assumes that over limited domains atmospheric motions may be treated as incompressible.

- Pressure fluctuations are small in the sense $\frac{p'}{p} \ll \frac{T'}{T} \ll 1$. From the ideal gas law,

$$\frac{\rho'}{\overline{\rho}} \approx -\frac{T'}{\overline{T}} \approx -\frac{\theta'}{\overline{\theta}}. \tag{9.12}$$

θ is either the potential temperature or the virtual potential temperature.

- Fluctuations in density are only important to the extent that they lead to fluctuations in the buoyancy. ρ may, therefore, be replaced by $\overline{\rho}$, except when it occurs in the product ρg.

- Density does not vary along streamlines,

$$\frac{d\rho}{dt} = \frac{d\overline{\rho}}{dt} = 0. \tag{9.13}$$

From equations (9.13) and (F.1), the equation of continuity is

$$\frac{\partial u_i}{\partial x_i} = 0. \tag{9.14}$$

If we write $u = \overline{u} + u'$ and average (9.14), we find,

$$\frac{\partial \overline{u}_i}{\partial x_i} = \frac{\partial u'_i}{\partial x_i} = 0. \tag{9.15}$$

9.1.5 Eddy energy

We use the inertial equations, (F.4), the stress equation, (F.5), and decompose u_i, p, ρ, and τ_{ij}^m into mean and fluctuating components:

$$\begin{aligned}
\frac{du_i}{dt} &= \frac{\partial \overline{u}_i}{\partial t} + \frac{\partial u'_i}{\partial t} + \overline{u}_j \frac{\partial \overline{u}_i}{\partial x_j} + u'_j \frac{\partial u'_i}{\partial x_j} + \overline{u}_j \frac{\partial u'_i}{\partial x_j} + u'_j \frac{\partial \overline{u}_i}{\partial x_j}, \\
&\approx -g_i - \frac{1}{\overline{\rho}} \frac{\partial \overline{p}}{\partial x_i} + \frac{1}{\overline{\rho}} \frac{\partial \overline{\tau}_{ij}^m}{\partial x_j} + g_i \frac{\theta'}{\overline{\theta}} - \frac{1}{\overline{\rho}} \frac{\partial p'}{\partial x_i} + \frac{1}{\overline{\rho}} \frac{\partial (\tau_{ij}^m)'}{\partial x_j}. \tag{9.16}
\end{aligned}$$

The Boussinesq approximations have been used, and we have divided throughout by $\overline{\rho}$.

Averaging over (9.16), all terms containing a single primed quantity vanish and

$$\begin{aligned}
\overline{\frac{du_i}{dt}} &= \frac{\partial \overline{u}_i}{\partial t} + \overline{u}_j \frac{\partial \overline{u}_i}{\partial x_j} + \overline{u'_j \frac{\partial u'_i}{\partial x_j}}, \\
&= -g_i - \frac{1}{\overline{\rho}} \frac{\partial \overline{p}}{\partial x_i} + \frac{1}{\overline{\rho}} \frac{\partial \overline{\tau}_{ij}^m}{\partial x_j}. \tag{9.17}
\end{aligned}$$

If we now subtract (9.17) from (9.16) and use (9.15), we obtain an equation for the eddy components:

$$\frac{\partial u'_i}{\partial t}+\overline{u}_j\frac{\partial u'_i}{\partial x_j}+u'_j\frac{\partial \overline{u}_i}{\partial x_j}+\frac{\partial(u'_iu'_j)}{\partial x_j}-\overline{u'_j\frac{\partial u'_i}{\partial x_j}} = g_i\frac{\theta'}{\overline{\theta}}-\frac{1}{\overline{\rho}}\frac{\partial p'}{\partial x_i}+\frac{1}{\overline{\rho}}\frac{\partial(\tau_{ij}^m)'}{\partial x_j} . \quad (9.18)$$

We now form an equation for the eddy kinetic energy, equation (9.10), by multiplying (9.18) by u'_i and averaging. After some manipulation, interchange of indices, and use of equation (9.15),

$$\frac{d\overline{k^e}}{dt} = \underbrace{\frac{\tau_{ij}^e}{\overline{\rho}}\frac{\partial \overline{u}_i}{\partial x_j}}_{shear}-\underbrace{\frac{\partial}{\partial x_i}\overline{u'_i\left(\frac{p'}{\overline{\rho}}+k^e\right)}}_{transport}+\underbrace{g_i\frac{\overline{u'_i\theta'}}{\overline{\theta}}}_{buoyancy}+\underbrace{\overline{\frac{u'_i}{\overline{\rho}}\frac{\partial(\tau_{ij}^m)'}{\partial x_j}}}_{dissipation} . \quad (9.19)$$

In (9.19) the complete differential on the left is formed using mean velocities rather than actual velocities (i.e., mean + eddies); this leads to the appearance of a kinetic energy term in the second term on the right side. This is the only possible procedure if the eddy motions are not directly observed, and their presence is inferred only from the fluxes that they cause.

The left side of (9.19) is the rate of change of the average eddy kinetic energy; the terms on the right are sources and sinks. To the extent that there is any analogy between eddy stresses and viscous stresses (we shall return to this topic), the inequality (2.5) should hold for the first term, and *shear generation* is almost always positive. The second term is the divergence of the eddy flux of eddy energy, with an additional work term that combines with eddy energy to form an eddy enthalpy. The eddy flux of eddy enthalpy causes *transport of eddy energy* from one part of the system to another.

The third term on the right involves *buoyancy* associated with the thermal expansion of air. Because $g_i = (0,0,g)$, this term may be written

$$g_i\frac{\overline{u'_i\theta'}}{\overline{\theta}} = g\frac{\overline{w's'}}{c_p} = \frac{g}{\overline{\rho}c_p}F_z^e(s) . \quad (9.20)$$

s is the entropy, and $F_z^e(s)$ is the vertical eddy flux of entropy. $F_z^e(s)$ is also approximately equal to $\frac{F_z^e(h)}{T}$ where h is the enthalpy. This follows from equation (2.21) because π cannot fluctuate at a fixed level. We have used the definition of potential temperature, equation (2.28), and the Boussinesq condition (9.12). Thus, buoyancy generates eddy energy if the entropy and enthalpy fluxes are upwards, and vice versa.

The fourth term on the right of (9.19) may be written approximately, for ν and $\overline{\rho}$ constant, and with the help of equation (F.6),

$$\overline{\frac{u'_i}{\overline{\rho}}\frac{\partial(\tau_{ij}^m)'}{\partial x_j}} = \nu\overline{u'_i\frac{\partial^2 u'_i}{\partial x_j^2}} = \nu\frac{\partial^2 \overline{k^e}}{\partial x_j^2}-\nu\overline{\left(\frac{\partial u'_i}{\partial x_j}\right)^2} . \quad (9.21)$$

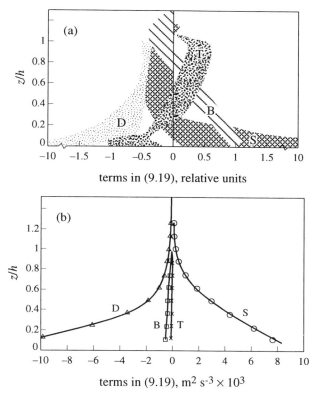

Figure 9.1 Sources and sinks of eddy energy. h is the height of the boundary layer. The terms on the right side of equation (9.19) are: D, dissipation; S, shear; B, buoyancy; T, transport. (a) Observations and model calculations for a daytime, convective boundary layer. The shaded areas show the range of data. (b) Clear, night-time conditions.

The first term on the right of (9.21) is small, while the second is equal to the irreversible work per unit volume performed by the eddies against molecular viscosity (see equation 2.5). Because molecular dissipation has little direct effect on large-scale motions, this term is effectively the entire irreversible work term, ϕ_{irr}.

The terms on the right of equation (9.19) are illustrated for both convective (i.e., thermally unstable, see Table 2.2) and nonconvective conditions in Figure 9.1. We shall return to these data in later sections, but the following comments may be made now. For the nonconvective conditions illustrated in Figure 9.1(b), shear generation and dissipation are the only significant terms, and these approximately balance. There is negligible transport from one part of the boundary layer to another, so that generation and dissipation balance locally. This is known as *local dissipation*.

For the convective case in Figure 9.1(a), the transport term is large, and generation is not in local balance with dissipation (*nonlocal dissipation*). Of the two generation terms, shear exceeds buoyancy below $z/h = 0.2$, but buoyancy dominates above this level. This transition is the basis for the *Monin-Obukhov similarity theory* (see §9.3.2).

9.1.6 The Ekman and the surface layers

A kilometer or two above the surface, in middle latitudes, the effect of the surface is felt only weakly and the wind is geostrophic, governed by equations (F.14) and (F.15). As we rise from the surface into the free atmosphere, the wind direction changes, and the wind speed increases with height. This change of wind with height is, under idealized conditions, known as the *Ekman spiral*, and the region in which it takes place is the *Ekman layer*.

We may examine the physics of the Ekman layer by adding a frictional term to equation (F.14) and writing the equation in terms of averaged velocities and averaged pressures. First we require an expression for the stress. From equations (F.4) and (F.5) we may show that the body force in equation (F.10) is, in terms of the stress, $f_x = \frac{1}{\rho}\frac{\partial \tau_{xi}}{\partial x_i}$. Following the ideas that we have developed in previous sections, if we wish to write (F.10) in terms of mean velocities and mean pressures, the stress must be interpreted as the eddy stress. We consider only the variation of the eddy stress in the vertical. The geostrophic equation (F.14) now becomes

$$0 \approx -\frac{1}{\bar{\rho}}\frac{\partial \bar{p}}{\partial x} + 2\Omega \bar{v} \sin \phi + \frac{1}{\bar{\rho}}\frac{\partial \tau_{xz}^e}{\partial z} . \qquad (9.22)$$

According to our earlier discussion, the second term on the right of (9.22) is small near to the ground and increases with height. On the other hand, the third term on the right is large near to the ground and decreases with height. The first term on the right is controlled by synoptic events, and we assume it to be constant in the boundary layer. Above a certain level, $z = z_e$, we expect the first and second terms on the right of (9.22) to balance and to give a mean geostrophic wind:

$$\bar{v}_g = \frac{1}{2\bar{\rho}\Omega \sin \phi}\frac{\partial \bar{p}}{\partial x} . \qquad (9.23)$$

In middle latitudes, $2\Omega \sin \phi \approx 10^{-4}$ s^{-1}, $\frac{1}{\bar{\rho}}\frac{\partial \bar{p}}{\partial x} \approx 10^{-3}$ m s^{-2}, and $\bar{v}_g \approx 10$ m s^{-1}.

A treatment of the Ekman spiral may be found in most meteorology texts. We require an analytical expression for the stress. The simplest treatment involves the technique of *K-closure*, which will be discussed in

§9.2.2. If we introduce an *eddy diffusion coefficient*, K^e (see equation 9.27), the third term on the right of equation (9.22) may be written

$$\frac{1}{\rho}\frac{\partial \tau^e_{xz}}{\partial z} \approx K^e \frac{\partial^2 \overline{u}}{\partial z^2}. \tag{9.24}$$

With this approximation (9.22) may be solved simultaneously with the zonal equation. Assuming $u \sim v$ and $\frac{\partial^2}{\partial z^2} \sim \frac{1}{z^2}$ it may be shown that the second and third terms in (9.22) are comparable when

$$z = z_e = \left(\frac{K^e}{\Omega \sin \phi}\right)^{1/2}. \tag{9.25}$$

z_e is the height of the Ekman layer. A typical value for the eddy coefficient, K^e, in the free atmosphere is 5 m² s⁻¹; with $2\Omega \sin \phi = 10^{-4}$ s⁻¹, $z_e \sim$ 300 m.

Equations (9.25) and (9.23) both contain $\sin \phi$ in the denominator, and have a singularity at the equator. This is the reason for reiterating the words "in middle latitudes" when dealing with Coriolis terms. Geostrophic theory and the theory of the Ekman layer are inapplicable in the tropics, but theories of the surface layer and the convective layer remain valid.

Turning now to the *surface or constant-stress layer*, we expect the balance near to the surface to be between the first and third terms on the right of (9.22). At the surface, the stress is finite and equal to τ_0. According to this balance, the stress will deviate from this surface value for heights greater than

$$z_s \approx \frac{\tau_0}{\partial \overline{p}/\partial x}. \tag{9.26}$$

z_s is the height of the surface layer. Below this height we assume that $\tau = \tau_0$, which greatly simplifies the equations. For $\tau_0 \sim 0.1$ kg m⁻¹ s⁻², and $\frac{\partial \overline{p}}{\partial x} \sim 10^{-3}$ kg m⁻² s⁻², $z_s \sim 100$ m.

9.2 Closure

9.2.1 The moment equations

Equation (9.17) expresses mean values of the velocity (first order in the velocity) in terms of averages over products of two eddy variables (second order in the velocity), while equation (9.19) expresses a second-order quantity in terms of one of third order. Equations can be constructed for a quantity of any order in terms of higher order quantities. *Closure techniques* involve, as the name implies, closing these equations with approximate expressions for a high-order quantity in terms of lower order quantities. *Local closure*, appropriate for local dissipation, employs relations between orders at the same location. Large computers make it possible to solve problems with high-order closure, from which it has been demonstrated that the higher the order of the closure the more accurate, in most cases, is the solution.

The lowest order closure is the second, which approximates second-order quantities, such as eddy stresses, in terms of local mean winds. This method, now known as K-closure, dates back to the first days of turbulence theory.

For the remainder of this chapter we shall restrict our attention to situations in which the wind does not change direction with height; for this situation we may simplify the boundary-layer equations by adopting axes such that $\overline{w} = \overline{v} = 0$. Without loss of generality we may also assume that the wind varies only with height, $\frac{\partial \overline{u}}{\partial x} = \frac{\partial \overline{u}}{\partial y} = 0$. According to these assumptions all fluxes must be vertical and we may omit z-suffixes.

9.2.2 K-closure

K-closure is based on an analogy with molecular diffusion. For molecular diffusion, fluxes are proportional to the negative of the gradient of the diffused quantity. The approximation adopted is

$$F^e(\chi) = -\overline{\rho} K^e \frac{\partial \overline{\chi}}{\partial z}, \qquad (9.27)$$

where K^e is called the *eddy diffusion coefficient*.

An early rationale for this closure was that eddies preserve their properties over an average displacement, l, the *mixing length*, and then mix with their surroundings. It follows that

$$\chi' \sim -l \frac{\partial \overline{\chi}}{\partial z} \qquad (9.28)$$

and, from the definition of eddy flux,

$$K^e \sim \overline{w'l} . \qquad (9.29)$$

The distance that an eddy can travel must correlate positively with its velocity, so that $K^e > 0$.

Equation (9.29) suggests that K^e should be the same for any transferable quantity that does not affect the motion of the eddy. This exempts only fluxes of thermal quantities, entropy, enthalpy etc., because we saw from the discussion of equations (9.19) and (9.20), that the magnitudes of these quantities can affect the dynamical state of the eddies.

If we recall that a stress is a negative momentum flux and that momentum per unit mass is velocity, we find from (9.2) and (9.27)

$$F^e(\overline{u}) = -\rho K^e \frac{\partial \overline{u}}{\partial z} = -\rho u_*^2 . \qquad (9.30)$$

THE PLANETARY BOUNDARY LAYER

For *neutral flows*, when $\frac{d\theta}{dz} = 0$ and the entropy flux is zero, Nikuradse's experimental result, (9.3), should be valid, and, combining (9.3) and (9.30),

$$K^e = k^2 z^2 \frac{\partial \overline{u}}{\partial z} . \tag{9.31}$$

Equations (9.31) and (9.27) have been used to predict boundary-layer fluxes of species such as water vapor, under neutral conditions, given direct measurements of $\frac{\partial \overline{u}}{\partial z}$ and $\frac{\partial \overline{X}}{\partial z}$ only, a simple but useful result.

9.2.3 Mixing above the surface layer

Entropy and enthalpy

The theoretical underpinning for K-closure may be weak, but it is the only numerical technique at present available to represent mixing by processes whose scale is smaller than the grid size of a numerical calculation. The "mixing-length" rationale for K-closure suggests that the theory should be applied to quantities that are conserved over vertical displacements. If the eddy motions are rapid enough, the motions will be adiabatic and entropy and static energy qualify as transferable quantities. From equations (9.27) and (2.28),

$$F^e(s) = -K_H^e \overline{\rho} \frac{\partial \overline{s}}{\partial z} = -\frac{K_H^e \overline{\rho} c_p}{\overline{\theta}} \frac{\partial \overline{\theta}}{\partial z} . \tag{9.32}$$

The subscript H (for "heat") has been used as a reminder that eddy diffusion coefficients for thermal functions may differ from those for momentum. We may now sharpen the statements made in §5.1.2 concerning relationships between observed lapse rates and the adiabatic lapse rate. According to (9.32), the entropy flux is upwards if $\frac{\partial \overline{\theta}}{\partial z} < 0$ (this is an unstable atmosphere; see Table 2.2) and, according to (9.19) and (9.20), this will cause the eddy energy to increase; it should also cause K_H^e to increase, because according to (9.29), K_H^e increases with eddy amplitudes. This creates a positive feedback, and the eddy flux could increase indefinitely. However, eddy fluxes are limited by external conditions and cannot increase indefinitely. The alternative way to limit the flux is for $\frac{\partial \overline{\theta}}{\partial z} \to 0$, or for $\Gamma \to \Gamma_{ad}$, and this is what happens.

Species

The conserved quantity is mixing ratio, $m_i = \rho_i/\rho$. The flux of species, i, is

$$\overline{w' \rho_i'} = \overline{w'(\rho m_i)'} = F^e(m_i) = -K^e \overline{\rho} \frac{\partial \overline{m_i}}{\partial z} . \tag{9.33}$$

If we now define a *density scale height* by

$$\frac{\partial \overline{\rho}}{\partial z} = -\frac{\overline{\rho}}{H_\rho},\qquad(9.34)$$

then,

$$F^e(m_i) = -K^e\left(\frac{\partial \overline{\rho_i}}{\partial z} - \frac{\overline{\rho_i}}{\overline{\rho}}\frac{\partial \overline{\rho}}{\partial z}\right) = \overline{\rho_i}K^e\left(\frac{1}{H_{\rho,i}} - \frac{1}{H_\rho}\right),\qquad(9.35)$$

where $H_{\rho,i}$ refers to the species and H_ρ to air. The direction of the flux depends upon the relative magnitudes of these two scale heights.

Diffusion times

Eddy diffusion times are important for understanding the distribution of species, including ozone, as in Figure 6.7. In the absence of ozone production or destruction, equation (6.37) may be combined with (9.33) to give

$$\frac{\partial \overline{m_i}\overline{\rho}}{\partial t} - \frac{\partial}{\partial z}K^e\overline{\rho}\frac{\partial \overline{m_i}}{\partial z} \approx 0.\qquad(9.36)$$

For vertical transport we may assume that m_i varies significantly over a scale height, $\frac{\partial}{\partial z} \sim \frac{1}{H}$, and that $\frac{\partial}{\partial t} \sim \frac{1}{\tau_e}$, where τ_e is the *eddy diffusion time* (not to be confused with the eddy stress). With these substitutions,

$$\tau_e \sim \frac{H^2}{K^e},\qquad(9.37)$$

in agreement with equation (1.8), but for the different diffusion coefficient.

9.3 Similarity

For local dissipation, all parameters, steady and fluctuating, should be related to others at the same location. If the interrelated parameters can be identified, dimensional analysis is a powerful tool. *Similarity theories* require that profiles of dimensionless atmospheric parameters must have a unique form when stated in terms of relevant independent variables, also in dimensionless form. The form or profile may be determined empirically once this parametric dependence has been established.

9.3.1 The Richardson number

A dimensionless parameter of crucial importance for this discussion is the *flux Richardson number*, Rf. This number is the ratio of the buoyancy to the shear terms in the eddy energy equation, (9.19). With the help of equation (9.20) we find, in the constant-stress layer,

$$Rf = -\frac{gF^e(s)}{c_p\tau_0 \partial \overline{u}/\partial z} \ . \tag{9.38}$$

The denominator in (9.38) is almost always positive. The entropy flux is normally positive during the day but may be negative at night. In principle, either sign is possible for both generation terms, and either one may be numerically larger than the other. Thus the Richardson number may be either positive or negative and may be either large or small; the four possibilities represent different physical situations.

The negative sign in equation (9.38) was chosen by Richardson because he wanted to investigate the generation of shear turbulence in a stable atmosphere ($F^e(s) < 0$). Under this circumstance a balance between generation and destruction is possible when the Richardson number is of the order of magnitude of unity. More precisely, observations show that balance occurs when

$$Rf = Rf_c \approx 0.2 \ . \tag{9.39}$$

Rf_c is the *critical Richardson number*.[1] This balance is not a common circumstance in the boundary layer, as may be seen by inspection on Figure 9.1. In Figure 9.1(a) there is a narrow region near $z/h = 0.4$ where buoyancy and shear terms are equal and opposite. However, it is more common for one or the other to dominate, as in Figure 9.1(b), and also in Figure 9.1(a) for $z/h < 0.1$.

If $|R_f| \ll 1$, shear dominates and buoyancy is relatively unimportant. The sign is now more or less irrelevant because shear generation must be positive if the air is turbulent. Nikuradse's result equation (9.3) may now be used and the Richardson number written in the form,

$$Rf = -\frac{gkzF^e(s)}{\overline{\rho}c_p u_{*,0}^3}, \ |Rf| \ll 1 \ . \tag{9.40}$$

We now return to Figure 9.1. For the clear night-time conditions, equation (9.40) should apply, although it may not be particularly useful. Figure 9.1(b) is more interesting. At the lowest levels eddies are generated by shear flow. Buoyancy creation is positive but smaller: Rf is negative but numerically small. As height increases, so, according to (9.40), does the negative Richardson number until, from (9.38), buoyancy will begin to dominate. This transition height is called the *Monin-Obukhov length*, L. With $Rf = 1$,

$$z = L = -\frac{u_{*,o}^3 \overline{\rho} c_p}{kgF_z^e(s)} \ . \tag{9.41}$$

---

[1] This discussion is usually carried out in terms of another Richardson number, Ri, formed from Rf by substituting approximations for the two fluxes; in an attempt to minimize confusion, only the flux Richardson number is discussed here.

From the data in Table 9.1, $|L| \sim 30$ m but, depending upon the size of the entropy flux, it may be anywhere from 10 m to ∞.

In the next section we discuss the role of the Monin-Obukhov length in boundary-layer dynamics, where L and Rf can have either sign. The remaining circumstance is free convection, for which buoyancy dominates, and $|Rf| \gg 1$; this is the topic of §9.4.

9.3.2 Monin-Obukhov similarity

The similarity arguments that follow are not rigorous. They rest upon plausibility and ultimately on the test of observations. We seek a relevant, dimensionless, dependent variable that defines the wind profile, and assert that it should be a unique function of the relevant, independent variables. The discussion at the end of the previous section suggests that the independent height variable should be

$$\zeta = \frac{z}{L} . \tag{9.42}$$

This variable appears to combine all of the information that we have found to be important when $|Rf|$ is neither very small nor very large. For a dimensionless, dependent variable we look to Nikuradse's relation, equation (9.3). This equation is certainly correct for $|Rf| \ll 1$, and our objective is to extend its validity to include buoyancy effects. The Monin-Obukhov similarity hypothesis states that

$$\frac{kz}{u_{*,0}} \frac{\partial \overline{u}}{\partial z} = \phi_M(\zeta) , \tag{9.43}$$

where $k \approx 0.41$ is von Karman's constant. Because $\zeta = 0$ implies $F^e(s) = 0$, that is, neutral flow, it follows that

$$\phi_M(0) = 1 . \tag{9.44}$$

Let us review the status of equation (9.43). It is simple. It is dimensionally correct. It appears to involve all of the parameters that are important to the problem. It embodies Nikuradse's results. It is the only expression to be proposed that has all of these properties. It tells us nothing about the form of $\phi_M(\zeta)$, although additional arguments have been developed that restrict its form. Most important is the empirical test in Figure 9.2 that covers a very wide range of atmospheric conditions. The small scatter about a universal curve shows that the Monin-Obukhov similarity is valid.

Figure 9.2 also shows the universal function, $\phi_H(\zeta)$ (H indicates a thermal parameter), which governs the potential temperature profile,

$$\frac{kz}{\theta_{*,0}} \frac{\partial \overline{\theta}}{\partial z} = \phi_H(\zeta) . \tag{9.45}$$

THE PLANETARY BOUNDARY LAYER

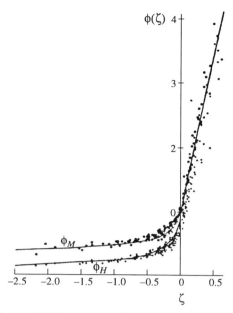

Figure 9.2 The universal functions $\phi_M(\zeta)$ and $\phi_H(\zeta)$.

$\theta_{*,0}$ is defined by

$$-\theta_{*,0} = \frac{\overline{(w'\theta')}}{u_{*,0}}, \qquad (9.46)$$

where

$$\overline{(w'\theta')} = \frac{\overline{\theta} F^e(s)}{\overline{\rho} c_p}. \qquad (9.47)$$

The choice of dimensionless variables for the potential temperature profile is reached by essentially the same reasoning that we used for the velocity profile. The data in Figure 9.2 demonstrate that the similarity hypothesis is as valid for temperature as it is for velocity but that the similarity function, $\phi(\zeta)$, differs for unstable atmospheres ($\zeta < 0$).

9.4 The convective boundary layer

9.4.1 Structure

When the transport term in equation (9.19) is finite, local similarity is inapplicable and alternate approaches are required. We shall examine one of the simplest of these, an integral treatment of a dry, diurnally driven, convective boundary layer over a land surface. A moist, marine boundary

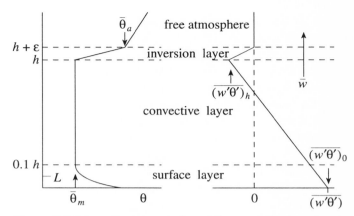

Figure 9.3 Schematic representation of the convective boundary layer.

layer, containing stratocumulus or stratus clouds, was mentioned briefly in §8.6.2. This is a more complicated situation, but it has been treated by analogous methods.

On a clear night, a shallow, temperature inversion (i.e., a stable region) forms at the earth's surface. When the sun starts to heat the ground, it must first overcome this inversion; thereafter, convective activity is forced upwards into the free atmosphere, which is usually slightly stable. As the convective layer thickens, it forms at its top a sharp interface or *capping inversion*, defining unambiguously the height of the boundary layer, h.[2] By midafternoon, the height of the convective layer will be 1 to 2 km, more over a desert, and less over water. A schematic representation of a developed, *convective boundary layer* is shown in Figure 9.3.

9.4.2 Assumptions

A description of Figure 9.3 will illustrate some of the implicit assumptions involved in this treatment. We treat the convective layer as a single entity. It is assumed that the convection is so intense that the layer is adiabatic and may be specified by a single potential temperature, $\bar{\theta}_m$. This assumption and the other assumptions we shall make are based on experience and observation.

The base of the convective layer is near to $0.1\ h$. This is greater than the Monin-Obukhov scale, and shear generation may be neglected. The

---

[2] We did not hesitate to talk of the height of the boundary layer in previous sections, but we were referring to a general vertical scale for the phenomenon. Here, by way of contrast, we are talking about a well-defined and measurable lid to a region of strong convection.

inversion layer is shown with a finite thickness, ϵ, which will be allowed to tend to zero.

Above the convective layer is the free atmosphere, with a stable lapse rate of potential temperature, $\left(\frac{\partial \overline{\theta}}{\partial z}\right)_a > 0$. In the stable, free atmosphere, all eddy quantities are assumed to be zero. A mean vertical velocity, $\overline{w}$, is assumed to exist throughout the inversion layer, forced by the ambient synoptic conditions.

There are some simple consequences of these assumptions. To examine them we assume that the vertical flow is adiabatic and write an equation for the conservation of entropy, with $\frac{d\theta}{dt} = 0$. The equation that we shall use is

$$\frac{\partial \overline{\theta}}{\partial t} + \overline{w}\frac{\partial \overline{\theta}}{\partial z} + \frac{\partial \overline{w'\theta'}}{\partial z} = 0 \,. \tag{9.48}$$

Equation (9.48) was obtained from equations (F.20) and (F.21) by writing $\xi = s = c_p \ln \theta$, allowing vertical variations only, and making the decomposition, (9.5), into mean and eddy components.

Equation (9.48) is valid both above and below the capping inversion, but with different constraints. Above the inversion, fluctuating quantities are zero and

$$\frac{\partial \overline{\theta}_a}{\partial t} = -\overline{w}\left(\frac{\partial \overline{\theta}}{\partial z}\right)_a \,. \tag{9.49}$$

In the convective layer, on the other hand, $\overline{\theta}$ is constant and equal to $\overline{\theta}_m$ so that $\frac{\partial \overline{w'\theta'}}{\partial z}$ is independent of height. The flux of entropy, therefore, varies linearly with height,

$$(\overline{w'\theta'})_z = (\overline{w'\theta'})_0 + \frac{z}{h}\left\{(\overline{w'\theta'})_h - (\overline{w'\theta'})_0\right\}, \text{ for } z \leq h \,. \tag{9.50}$$

9.4.3 The capping inversion

The behavior of this model is dominated by the two entropy fluxes $(\overline{w'\theta'})_0$ and $(\overline{w'\theta'})_h$. The former may be regarded as a given, external parameter connected to the insolation, but the *entrainment flux*, $(\overline{w'\theta'})_h$, is a dependent quantity and must be determined. First, we establish two relationships between the magnitude of the capping inversion, $\Delta\theta = (\overline{\theta}_a - \overline{\theta}_m)$, the entrainment flux, and the rate of change of the height of the inversion layer.

Integrating (9.48) from h to $h + \epsilon$,

$$\int_h^{h+\epsilon} \frac{\partial \overline{\theta}}{\partial t} dz + \overline{w}\Delta\theta + (\overline{w'\theta'})_a - (\overline{w'\theta'})_h = 0 \,. \tag{9.51}$$

Let $\epsilon \to 0$ in the identity,

$$\frac{\partial}{\partial t} \int_h^{h+\epsilon} \overline{\theta}\, dz = \int_h^{h+\epsilon} \frac{\partial \overline{\theta}}{\partial t}\, dz + \Delta\theta \frac{\partial h}{\partial t}. \qquad (9.52)$$

The left side of (9.52) tends to zero and, with $\overline{(w'\theta')}_a = 0$ in equation (9.51),

$$\overline{(w'\theta')}_h = -\Delta\theta \left(\frac{\partial h}{\partial t} - \overline{w} \right). \qquad (9.53)$$

The inversion layer rises through the early part of the day. At, but immediately above the inversion layer we may write, with the help of equation (9.49),

$$\begin{aligned}
\frac{d\overline{\theta}_a}{dt} &= \frac{\partial \overline{\theta}_a}{\partial t} + \left(\frac{\partial \overline{\theta}}{\partial z}\right)_a \frac{\partial h}{\partial t}, \\
&= \left(\frac{\partial \overline{\theta}}{\partial z}\right)_a \left(\frac{\partial h}{\partial t} - \overline{w}\right).
\end{aligned} \qquad (9.54)$$

Because $\overline{\theta}_m$ does not vary with height,

$$\frac{\partial \overline{\theta}_m}{\partial t} = \frac{d\overline{\theta}_m}{dt} = \frac{d\overline{\theta}_a}{dt} - \frac{d\Delta\theta}{dt}. \qquad (9.55)$$

Eliminating $\frac{d\overline{\theta}_a}{dt}$ between (9.55) and (9.45), and using (9.48) and (9.50),

$$\left(\frac{\partial \overline{\theta}}{\partial z}\right)_a \left(\frac{\partial h}{\partial t} - \overline{w}\right) - \frac{d\Delta\theta}{dt} = \frac{1}{h} \left\{ \overline{(w'\theta')}_0 - \overline{(w'\theta')}_h \right\}. \qquad (9.56)$$

Equations (9.53) and (9.56) describe the behavior of the convective layer. Each equation contains three unknowns, $\Delta\theta$, h, and $\overline{(w'\theta')}_h$. One additional item of information is required to obtain a solution.

9.4.4 Entrainment fluxes

To close the foregoing equations we need an expression for the entrainment flux. Entrainment is caused by buoyant eddies interacting with the inversion, and mixing stable air from above the inversion down into the convective layer. This requires work to be done by the eddies to break down the stable air and, from our discussion of equation (9.19), implies a negative entrainment flux. The drive for the convective layer is the convective flux at the base of the convective layer. This flux is positive and generates eddy energy. We may reasonably anticipate from this discussion that the two fluxes are related and write

$$\frac{\overline{(w'\theta')}_h}{\overline{(w'\theta')}_0} = -\beta. \qquad (9.57)$$

THE PLANETARY BOUNDARY LAYER

On the basis of observations $\beta \approx 0.2$.

Equations (9.53), (9.56), and (9.57) have a solution for the case $\overline{w} = 0$, which may be confirmed by substitution:

$$\Delta\theta = \frac{\beta}{1+2\beta} h \left(\frac{\partial\overline{\theta}}{\partial z}\right)_a , \qquad (9.58)$$

$$\frac{\partial h}{\partial t} = \frac{(1+2\beta)(\overline{w'\theta'})_0}{h(\partial\overline{\theta}/\partial z)_a} . \qquad (9.59)$$

Equations (9.58) and (9.59) describe the behavior of a dry, convective boundary layer when there is no mean vertical motion. With the following data, $\beta = 0.2$, $\left(\frac{\partial\overline{\theta}}{\partial z}\right)_a = 5$ K km^{-1}, $h = 1000$ m, and $\overline{w'\theta'}_0 = 0.2$ m s^{-1}K (equivalent to an energy flux of 260 W m^{-2}), they give $\Delta\theta = 0.7$ K and $\frac{\partial h}{\partial t} = 5.6 \times 10^{-2}$ m s^{-1}. If the surface entropy flux is constant, equation (9.59) may be integrated to give

$$h^2 = 2\frac{(1+2\beta)(\overline{w'\theta'})_0}{(\partial\overline{\theta}/\partial z)_a} t . \qquad (9.60)$$

If $h = 10^3$ m, $t = 2.4$ h.

All of these numbers are typical of a diurnal, convective boundary layer over land, and the theory appears to be correct, as far as it goes. At the very least it provides a framework for describing a complex and important atmospheric phenomenon, and features may be added to extend its validity to the marine boundary layer and to other interesting circumstances. It stands as an example of a practical approach to an exceptionally difficult problem, making the best use of observations in a framework of our knowledge of physical process.

9.5 Reading

Modern texts offer unified approaches to the boundary layer, in which the beginner may have difficulty discerning the fundamental (and fundamentally simple) physical ideas. For this reason it may be helpful to read an elementary, and outdated, but still useful monograph

Priestley, C.H.B., 1959, *Turbulent transfer in the lower atmosphere.* Chicago: University of Chicago Press.

A recent monograph is

Garratt, J.R., 1992, *The atmospheric boundary layer.* Cambridge: Cambridge University Press.

A more formal approach, covering essentially the same ground is

Sorbjan, Z., 1989, *Structure of the atmospheric boundary layer.* Englewood Cliffs, NJ: Prentice Hall.

An important book that emphasizes diffusion in the boundary layer is

Pasquill, F., and Smith, F.B., 1983, *Atmospheric diffusion*, 3rd ed. New York: Wiley.

9.6 Problems

9.1 **Monin-Obukhov profiles.**

(i) For $\zeta > 0$ (stable conditions), the similarity function, $\phi(\zeta)$, is equal to $1 + 5\zeta$ for all transferable quantities (see Figure 9.2). If the Monin-Obukhov length, L, is 100 m and the friction velocity, $u_{*,0}$, is 0.3 m s^{-1}, what is the velocity difference between the 200 m and 1 m levels?

(ii) If the potential temperature is 300 K, what is the temperature scale, $\theta_{*,0}$?

(iii) What is the difference in potential temperature between the 200 m and the 1 m levels?

9.2 **A slightly stable wind profile.** The data in Table 9.2 were recorded during the night, close to the surface.

Table 9.2 Wind and temperature data

Height m	Wind speed m s^{-1}	Potential temperature °C
4	5.50	30.22
2	4.74	30.17

(i) Obtain an expression for the Monin-Obukhov length in terms of the observed data.

(ii) Hence calculate numerical values for $u_{*,0}$, $\theta_{*,0}$, τ_0, $F_z^e(h_{\text{dry}})$, and L.

9.3 **Bulk transfer coefficients.** Numerical models of the weather and climate cannot usually afford the luxury of a detailed description of the boundary layer. Instead, it is common practice to assume that the fluxes of momentum, enthalpy, and water vapor may be related to the average properties of the atmosphere at the surface (subscript s) and at a reference level near the surface (subscript r). Bulk transfer coefficients for "drag" and for "heat", C_D and C_H, respectively, are defined by the following relationships:

$$\tau_0 = -\rho C_D \overline{u}_r (\overline{u}_s - \overline{u}_r),$$
$$F_z^e(h_{dry}) = \rho c_p \overline{u}_r C_H (\overline{T}_s - \overline{T}_r).$$

Show that, for $\zeta > 0$, $C_M = C_D$.

9.4 The convective boundary layer. The essence of the problem of the convective boundary layer, as discussed in the text, is to find an additional equation, based upon physical realities, to close equations (9.53) and (9.56). An extreme closure assumption is

$$\overline{(w'\theta')}_h = 0.$$

With this assumption obtain solutions for $\Delta \theta$ and h at time t in terms of their values at $t = 0$. Discuss the meaning of the limit $\overline{w} = 0$.

9.5 Priestley's theory of free convection. If the surface stress $u_{*,0} \to 0$, while the entropy flux remains finite and positive, we have $L \to 0$ and $\zeta \to -\infty$. Eddies are now generated solely by buoyancy forces, and we have a state of free convection. $u_{*,0}$ is now an irrelevant parameter and the Monin-Obukhov scaling argument becomes irrelevant. It is believed that in a free-convecting regime the mean thermal parameters that involve temperature, length, and time only are the eddy flux, $\overline{(w'\theta')}_0$, the temperature gradient, $\frac{\partial \overline{\theta}}{\partial z}$, the buoyancy parameter, $\frac{g}{\overline{\theta}}$, which multiplies temperature fluctuations to yield a net force, and the height, z.

Show that

$$\frac{\partial \overline{\theta}}{\partial z} = -\text{const.} \overline{(w'\theta')}_0^{2/3} \left(\frac{g}{\overline{\theta}}\right)^{-1/3} z^{-4/3}$$

and that

$$\phi_H(\zeta) \to \text{const.}(-\zeta)^{-1/3} \text{ as } \zeta \to -\infty.$$

9.6 Eddy diffusion coefficients. For neutral shear flow, the parameters governing the mean flow are the surface stress, the distance from the surface, and the fluid density.

(i) Use a dimensional argument to find the form of the eddy diffusion coefficient for neutral shear flow.

(ii) Determine the mean wind profile for this condition.

9.7 **The Bowen ratio.** The *Bowen ratio* is defined to be the ratio of the flux of dry enthalpy to the flux of latent heat,

$$Bo = \frac{\rho c_p \overline{w'T'}}{\rho l \overline{w'm_v'}}.$$

(i) Assuming a steady state, with no net heat flux into the ground, what is the evaporation rate in terms of the net radiative heat flux and the Bowen ratio?

(ii) Show that, for stable conditions,

$$Bo = \frac{c_p(T_1 - T_2)}{l(m_{v,1} - m_{v,2})},$$

where the subscripts 1 and 2 represent two levels close to the ground.

SOLUTIONS TO PROBLEMS

1.1 Barometric law
(i) (a) $p_\infty = 3.6 \times 10^{-1}$ Pa. (b) $p_\infty \approx 10^{-153}$ Pa.
(ii) $p_\infty = 8.5 \times 10^{-13}$ Pa.

1.2 The critical level
(i) $p_e = \frac{mg}{\pi\sigma^2}$.
(iii) $p_e = 1.13 \times 10^{-6}$ Pa. $z_e \approx 430$ km.

1.3 Jeans escape
$\sim 10^{45}$.

1.4 Local thermodynamic equilibrium
(i) 2×10^{-20} J molecule^{-1}.
(ii) 4×10^{-21} J molecule^{-1}.
(iii) As high as the translational levels are in equilibrium, ~ 500 km.

1.5 Photodissociation

Absorber	λ_{diss}, nm	z_{diss}, km
H_2	277.1	~ 35
N_2	127.1	80-120
O_2	242.5	~ 35
O_3	1179.6	Ground level
$ClNO_2$	1098.1	Ground level
CO_2	277.5	~ 30

2.1 Margules' calculations
(i) $\overline{h}_{\text{dry}} = \frac{c_p \theta}{1 + r/c_p}\left\{\left(\frac{p_1}{p_0}\right)^{1+r/c_p} - \left(\frac{p_2}{p_0}\right)^{1+r/c_p}\right\}$.
(ii) $\Delta \overline{h}_{\text{dry}} = 13.55(\theta_2 - \theta_1)$ J kg^{-1}.
(iii) $\Delta \overline{h}_{\text{dry}} = 70.2(\theta_2 - \theta_1)$ J kg^{-1}.
(iv) 16.5 and 37.5 m s^{-1}, respectively.

2.2 The tephigram

(iii) $\Delta e = c_v(T_2 - T_1) = -7.18 \times 10^3$ J kg^{-1}. $\Delta h = c_p(T_2 - T_1) = -1.006 \times 10^4$ J kg^{-1}. $\Delta s = c_p \ln \frac{T_2}{T_1} - r \ln \frac{p_2}{p_1} = +10.45$ J kg^{-1}. $\Delta q_{\text{dry}} = \frac{1}{2}(T_2+T_1)\Delta s = 2.905 \times 10^3$ J kg^{-1}. $\Delta w = \Delta e - \Delta q = -10.09 \times 10^{-3}$ J kg^{-1}.

Positive signs indicate that the interaction is performed *on* the parcel; negative signs indicate that it is performed *on* the environment.

2.3 Moist and dry adiabats

(i) Trajectory I: θ_d = constant = T_0 = 288 K. Trajectories II and III (no difference): $\theta_d = 288 \exp\left\{\frac{lm_v(T_0)}{c_{p,a}T_0} - \frac{lm_v(T)}{c_{p,a}T}\right\}$.

(ii) $m_v(T_0) = \frac{p_e(T_0)}{p_0} \frac{r_a}{r_v} = 1.06 \times 10^{-2}$. $\Delta\theta_d = 27.6$ K

2.4 Moist entropy

20°C: $m_v = 1.454 \times 10^{-2}$; $l = 2.454 \times 10^6$ J kg^{-1}.
0°C: $m_v = 3.800 \times 10^{-3}$; $l = 2.501 \times 10^6$ J kg^{-1}.

$$\Delta s = \underbrace{(1-m)c_{p,a} \ln \frac{T_1}{T_2}}_{\text{term 1}} + \underbrace{mc \ln \frac{T_1}{T_2}}_{\text{term 2}} + \underbrace{\frac{m_v(T_1)l(T_1)}{T_1}}_{\text{term 3}} - \underbrace{\frac{m_v(T_2)l(T_2)}{T_2}}_{\text{term 4}}.$$

Table of numbers

Term (J kg^{-1}K^{-1})	Exact	Approximations		
		(i)	(ii)	(iii)
1	7.009×10	7.009×10	7.009×10	7.112×10
2	4.335	4.335	0	0
3	1.215×10^2	1.240×10^2	1.215×10^2	1.215×10^2
4	3.479×10	3.479×10	3.479×10	3.479×10
Δs	1.611×10^2	1.636×10^2	1.568×10^2	1.578×10^2

2.5 Specific heat of saturated vapor

(iii) (a) -0.35 K h^{-1}. (b) -0.11 K h^{-1}.

2.6 A Föhn wind

Approximate calculations

Position	m	θ_d, K	T_c, K	θ_e, K
D	4×10^{-3}	311.16	265.85	323.02
A	10^{-2}	294.66	285.49	321.47
A'	5×10^{-3}	296.45	272.75	310.27

The origin of the air at D is more likely to be A than A'.

2.7 Altitude calculation

(ii) 4.52 and 3.57 km, respectively.

2.8 Mixing of saturated air masses

(ii) $T_1 + T_2 - 2T = -0.91$. This number is always negative for water vapor, which ensures condensation. The fraction condensed is $\sim 5\%$.

2.10 Solar energy

(ii) Use nonthermal conversion, for example, photochemical or photoelectric, that does not immediately thermalize an absorbed photon.

3.1 Equivalent widths

(i)
$$W(u) = \alpha_L \int_{-\infty}^{+\infty} \left\{ 1 - \exp\left(-\frac{2u}{x^2+1}\right) \right\} dx.$$

$2u$ is the optical path at the line center.

(ii) Weak line: $W(u) = Snl$. Strong line: $W(u) = 2\sqrt{\alpha_L Snl}$.

3.2 Detection of deuterium

(i) $\alpha_D(H) = 1.652 \times 10^{-3}$ nm. $\alpha_D(D) = 1.168 \times 10^{-3}$ nm.

(ii) Because the absorption is weak it will be linear in the absorption coefficient. D/H$= 1.49 \times 10^{-9}$.

(iii) To order of magnitude the absorption by H will be unchanged, while the peak absorption of D will decrease in the ratio $\pi^{1/2}\alpha_D(D)/0.01$. The minimum detectable D/H ratio becomes D/H$= 7.2 \times 10^{-9}$.

3.3 The random band model

(i) $\overline{A} = \frac{W}{\delta}$.

(ii) $\mathcal{T}(\nu = 0) = \prod_{i=1}^{N} \exp\{-Snlf(\nu_i)\}$.

$$\overline{\mathcal{T}} = \frac{\prod_{i=1}^{N} \int_{-\frac{N\delta}{2}}^{+\frac{N\delta}{2}} \exp\{-Snlf(\nu_i)\} \frac{d\nu_i}{\delta}}{\prod_{i=1}^{N} \int_{-\frac{N\delta}{2}}^{+\frac{N\delta}{2}} \frac{d\nu_i}{\delta}},$$

$$= \frac{\left[\int_{-\frac{N\delta}{2}}^{+\frac{N\delta}{2}} \exp\{-Snlf(\nu_i)\} \frac{d\nu_i}{\delta}\right]^N}{N^N},$$

$$\rightarrow \exp -\frac{W}{\delta} \text{ as } N \rightarrow \infty.$$

(iii) (i) and (ii) are the same if $\frac{W}{\delta} \ll 1$.

3.4 Once in a blue moon

(i) A blue color requires more scattering in the red than in the blue. For $\tilde{=}1.33$, this is possible if ρ is restricted to the range $6 < \rho < 12$.

(ii) To fit an octave of visible light into this range requires a very narrow distribution of aerosol particle sizes, with a mean radius of 2.0 µm. According to the discussion in Chapter 8, this is a very unusual distribution of aerosol particles.

3.5 Measurement of extinction

According to Figure 3.13, if the aperture in angular units, $\frac{r}{f}$, is less than $\sim \frac{4}{x}$, all scattered light will be excluded and $Q_e = 2$. For larger apertures the light in the diffraction peak will enter the aperture and will not be counted as scattered light; we will measure $Q_e = 1$. For $\lambda = 0.5$ µm the critical particle radius is ~ 160 µm, corresponding to a drizzle drop.

3.6 Atmospheric extinction

The solution may be reached iteratively. If done thoughtfully one iteration is enough. Number of molecules per square meter $= 2.13 \times 10^{29}$. Number of aerosol particles per square meter $= 5.30 \times 10^8$. Number of ozone molecules per square meter $= 8.26 \times 10^{22}$.

4.1 Thermal emission from an isothermal, nonblack cloud, I

(ii) See solution to Problem 4.2.

(iii) We require $a < 0.2$ before the cloud is anything close to "black." According to Figure 8.10 this will only be correct for $\lambda > 30$ µm.

4.2 Thermal emission from an isothermal, nonblack cloud, II

Thermal flux from a cloud

| | | $\frac{F(0)}{\pi B}$ | |
a	$g = 1.0$	$g = 0.5$	$g = 0.0$
1.0	1.000	0.000	0.000
0.8	1.000	0.746	0.635
0.6	1.000	0.870	0.788
0.4	1.000	0.933	0.881
0.2	1.000	0.973	0.948
0.0	1.000	1.000	1.000

When $g = +1$, scattering is forward only, which is equivalent to having no scattering at all, only absorption.

4.4 Solar reflectivity or cloud albedo

(i)

$$\frac{d^2 F}{d\tau^2} = \alpha^2 F + \frac{f_0 a}{\mu_\odot} e^{-\tau/\mu_\odot}, \ \alpha^2 = 3(1-a).$$

$$\frac{dF}{d\tau} = 2(1-a)F - a f_0, \text{ at } \tau = 0.$$

(ii)
$$F = Xe^{-\alpha\tau} + Ye^{-\tau/\mu_\odot},$$
$$Y = \frac{f_0 a \mu_\odot}{1 - \alpha^2 \mu_\odot^2},$$
$$X = \frac{af_0 - Y\left\{\frac{1}{\mu_\odot} + 2(1-a)\right\}}{\alpha + 2(1-a)},$$
$$A = \frac{a}{\{1 + \frac{2}{3}\sqrt{3(1-a)}\}\{1 + \mu_\odot \sqrt{3(1-a)}\}}.$$

(iii)

Solar albedo for $\mu_\odot = \frac{2}{3}$

a	A
1.0000	1.0000
0.9990	0.9299
0.9900	0.7956
0.9000	0.4829
0.7000	0.2627
0.0000	0.0000

4.5 Line formation in a cloudy atmosphere
(ii)
$$\frac{A_c - A_\nu}{A_c} \rightarrow \left(\frac{2}{3} + \mu_\odot\right)\sqrt{3\frac{k_{\nu,m} n_m}{s_p n_p}}, \text{ as } 1 - a_\nu \rightarrow 0.$$

For the case of a thin absorbing atmosphere without clouds, we would expect linear absorption, proportional to $k_{\nu,m} n_m$, rather than to the square root of the same quantity. It is possible to devise observational tests that can distinguish between these two cases.

4.6 Direct and diffuse solar fluxes
(i)
$$F_\odot(\tau) = -\mu_\odot f_0 e^{-\tau/\mu_\odot},$$
$$F_s(\tau) = f_0 a \mu_\odot e^{-\tau/\mu_\odot},$$
$$F_l(\tau) = -f_0 a \mu_\odot \sqrt{3(1-a)} \left(\frac{2}{3} + \mu_\odot\right) e^{-\alpha\tau}.$$

(ii) $F_s(0) = -0.9990 F_\odot(0)$. $F_l(0) = 7.30 \times 10^{-2} F_\odot(0)$. For practical purposes, the *net* flux at $\tau = 0$ is F_l. It is directed downwards.

(iii) $\tau = 1.81$.

(iv) At this level the net flux is equal to the direct solar flux, and is directed downwards. $F_\odot(\tau)/F_\odot(0) = 0.170$.

4.7 Invariant imbedding

(ii) Contributions to $\frac{\partial R(\tau;\mu,\mu_0)}{\partial \mu}$:

from A, $-\left(\frac{1}{\mu} + \frac{1}{\mu_0}\right) R(\tau;\mu,-\mu_0)$;

from B, $\frac{aP(\mu,-\mu_0)}{4\mu\mu_0}$;

from C, $\frac{a}{2\mu_0} \int_0^1 R(\tau;\mu,-\mu')P(-\mu',-\mu_0)\,d\mu'$;

from D, $\frac{a}{2\mu} \int_0^1 R(\tau;\mu',-\mu_0)P(\mu,\mu')d\mu'$;

from E, $a \int_0^1 R(\tau;\mu,-\mu') \int_0^1 R(\tau;\mu'',-\mu_0)P(-\mu',\mu'')\,d\mu'\,d\mu''$.

4.9 Cooling to space

(i) and (iii)

Cooling to space

Band, cm^{-1}	$\overline{B}$, W m^{-2}Hz^{-1}	$\frac{\partial T}{\partial t}$, K day^{-1}
667	1.522×10^{-12}	1.67
2349	1.096×10^{-15}	1.34×10^{-2}

(ii)

$$\frac{\partial T}{\partial t} = -\frac{2\pi \overline{B} \mu_{CO_2}}{m_a c_p} \sum_{\text{band}} S_{\text{line}},$$

where $\overline{B}$ is the Planck function at the center of the band, μ_{CO_2} is the molar mixing ratio, and m_a is the mass of an air molecule.

5.1 The carbon dioxide greenhouse

Area under the curve is now 79 W m^{-2} ster^{-1}. If the carbon dioxide band is filled in, the area would be 90 W m^{-2} ster^{-1}. From these figures we may conclude that the ground temperature now is 265.7 K, while without the carbon dioxide absorption, it would be 257.2 K. The carbon dioxide contribution to the greenhouse is 8.5 K.

5.2 Radiative equilibrium in a semi-grey atmosphere

The solution is discussed in Goody and Yung (1989), p. 393 et seq.

5.3 Radiative time constants

(i)
$$\tau_{rad} \sim \frac{Mc_p}{\sigma T_e^3}.$$

(ii) The right side of (5.31) may be written $\frac{\tau_{dyn}}{\tau_{rad}}$. When this ratio is small dynamics should control the temperature distribution and lead to a small value of x.

(iii)

Radiative time constants

Planet	τ_{rad}, s	Planet days
Mars	2.38×10^5	2.8
Earth	1.12×10^7	1.3×10^2
Venus	1.14×10^9	1.1×10^2

(iv) The final column in the above table shows the number of planet days in a radiative time constant. If this number is small the atmosphere can adjust radiatively during 1 d, and the temperature can change greatly in the course of a day. If the number is large the temperature will not respond. Thermal tides are weak on Earth. They should also be weak on Venus, but they could be strong on Mars.

5.4 Limb darkening or brightening

(ii) The isophotes increase in radiance towards the limb (r increasing), that is, they exhibit limb brightening if $\beta < 0$, which means that the temperature increases with height, and vice versa for limb darkening. According to Chapman theory, this statement applies to the level $\tau_\nu = 1$.

(iii) The band center will show limb brightening and the wings will show limb darkening.

5.5 Remote sounding of temperature, I

(i) The maximum is at level 12, where the temperature is 220.7 K.

(ii) $I_\nu^+(\xi = 1) = 4.667 \times 10^{-2}$ W m^{-2}(wavenumber)$^{-1}$.

(iii) The brightness temperature is 224.2 K. The first approximation would place this temperature at level 12, whereas it occurs between levels 13 and 14.

5.7 Limb soundings

(iv) An advantage of limb sounding is that the kernel function is usually narrower than the Chapman function, and the vertical resolution is, therefore, better. Further, the very long slant path allows very weak absorbers to be measured. The main disadvantage is that no data are available for altitudes less than z_0.

$\frac{z-z_0}{H}$	Erf $\left(\frac{z-z_0}{H}\right)^{1/2}$	$h_\nu(z)/h_\nu(z_0)$
0.00	0.0000	∞
0.125	0.3830	5.295
0.25	0.5205	3.502
0.50	0.7460	2.186
0.75	0.7795	1.421
1.00	0.8427	1.000
1.25	0.8827	0.718
1.50	0.9167	0.522
1.75	0.9387	0.382
2.00	0.9545	0.282
3.00	0.9857	0.087
4.00	0.9943	0.028

6.1 Ozone and skin damage

(i) Hazard indices: normal ozone $\sim 6 \times 10^{-15}$; depleted ozone $\sim 15 \times 10^{-15}$.

(ii) UV-A has no skin sensitivity; UV-C has no photons; UV-B is therefore the only spectral range of importance for medical reasons.

6.4 Ozone and atmospheric motions

(i) (a) All boxes have the same mixing ratio. (b) $m_j = m_{j,e}$. Neither result is affected by the strength or dimensions of the circulation.

(ii) In both cases $m_4 = m_3 = m_2$. (a) $m_2 = m_{2,e}$, $m_1 = m_{1,e}$. (b) $m_2 = m_1 = m_{1,e}$. All boxes are controlled by the tropical stratosphere.

(iii) $m_1 = 8.0 \times 10^{-6}$, $m_2 = 5.8 \times 10^{-6}$, $m_3 = 5.5 \times 10^{-6}$, $m_4 = 5.3 \times 10^{-6}$.

6.5 The Chapman reactions

(i) $k_2 = 8.6 \times 10^{-46}$ m^{6}s^{-1}, $k_3 = 2.2 \times 10^{-21}$ m^{3}s^{-1}, $J_2 = 9 \times 10^{-10}$ s^{-1}, $J_3 = 4.5 \times 10^{-3}$ s^{-1}, which lead to $[O_3]_e = 5.2 \times 10^{17}$ m^{-3}, $[O]_e = 6.9 \times 10^{15}$ m^{-3}.

(ii) $[O^1D]_e = 8.6 \times 10^8$ m^{-3}.

6.6 Odd-nitrogen catalysis

$[O_3]_e = 4.7 \times 10^{17}$ m^{-3}.

6.7 Odd-chlorine reactions

$[Cl] = 3.3 \times 10^{12}$ m^{-3}. Almost all chlorine is in the form of ClO.

6.8 The Antarctic ozone hole

$[ClO] = 2 \times 10^{15}$ m^{-3}.

7.1 Source of hydroxyl

(i) 0.74%.

(ii) 2.0×10^5 m^{-3}.

(iii) 2.24×10^{12} m^{-3} s^{-1}.

7.2 Lifetime of OH in the continental boundary layer
(i) 18.7%.
(ii) 0.31 s.
(iii) 7×10^{11} m^{-3}.

7.3 Measurement of OH
From CH_3CCl_3, 4.7×10^{11} m^{-3}. From $CHCl_3$, 2.7×10^{11} m^{-3}.

7.4 Ca and Sr cycles
(i) $(Sr/Ca)_{sea} = 5(Sr/Ca)_{rivers}$.
(ii) $\frac{\tau_{Sr}}{\tau_{Ca}} = \frac{(Sr/Ca)_{sea}}{(Sr/Ca)_{rivers}} = 5$.

7.5 ^{14}C from nuclear explosions
3.19 y and 42.1 y.

8.2 Radar reflectivities
(i) The range of x is from 6.3×10^{-3} to 0.21. Rayleigh scattering should be a good approximation.

$$\Sigma = \int_0^\infty r^6 n(r) \left\{ 4\pi \left(\frac{2\pi}{\lambda}\right)^4 \left|\frac{\tilde{m}^2 - 1}{\tilde{m}^2 + 2}\right|^2 \right\} dr.$$

This expression places great weight upon large particles.

(ii) The larger drops reflect 10^4 times more than the smaller drops.

(iii) Rainfall $\propto \int_0^\infty n(r) r^5 \, dr$. The size weighting is similar to that for Σ, which should provide a fair measure of rainfall.

8.3 The melting band
$\left|\frac{\tilde{m}^2-1}{\tilde{m}^2+2}\right|^2$ is 5.3 times larger for water than for ice. Step I: reflectivity increases because of melting. Step II: number density decreases because fall velocity is higher for a droplet then for a snowflake.

8.5 The clouds of Venus
(i) The temperature becomes too high for the H_2SO_4 vapor mixing ratio, formed above the clouds, to maintain saturation.

(ii) Equation (8.29) suggests that these are old drops, all of similar age. If true, this implies that Mode III drops have not grown from Mode II by condensation (confirmed by other evidence).

(iii) 1.6×10^8 m^{-3}.

(iv) To fall through half the cloud takes 11.1 d for Mode III and 2.8 y for Mode II. It is consistent to look upon Mode II particles as very old.

(v)
$$\Delta r_{III} = \frac{\overline{E}\tau r_{II}}{6}.$$

9.1 Monin-Obukhov profiles
(i) $\overline{u}_{200} - \overline{u}_1 = 11.2$ m s^{-1}.
(ii) $\theta_{*,0} = 3.36 \times 10^{-2}$ K.
(iii) $\overline{\theta}_{200} - \overline{\theta}_1 = 1.25$ K.

9.2 A slightly stable wind profile
(i)
$$L = \frac{1}{\ln(z_1/z_2)} \left\{ \frac{(u_1 - u_2)^2 \overline{\theta}}{g(\theta_1 - \theta_2)} - 5(z_1 - z_2) \right\}.$$

(ii) $L = 501$ m, $u_{*,0} = 0.44$ m s^{-1}, $\theta_{*,0} = 0.029$ K, $F^e(h_{\text{dry}}) = -16.6$ W m^{-2}, $\tau_0 = 0.25$ Pa.

9.4 The convective boundary layer

$$h(t) = h(0) + \overline{w}t,$$

$$\Delta\theta(t) - \Delta\theta(0) = -\frac{\overline{(w'\theta')}_0}{\overline{w}} \ln\left\{\frac{h(t)}{h(0)}\right\}.$$

9.6 Eddy diffusion coefficients
(i) $K = \text{const.} \sqrt{\frac{\tau_0}{\rho}}$.

(ii) From (9.27), $\frac{\partial \overline{u}}{\partial z} = \frac{\text{const.}}{z} \sqrt{\frac{\tau_0}{\rho}}$, leading to the profile (9.4).

9.7 The Bowen ratio
(i)
$$E = \frac{-F(\text{rad})}{lB_0}.$$

Appendix A

NUMBERS AND UNITS

General constants
  Stefan-Boltzmann constant, $\sigma = 5.67032 \times 10^{-8}$ W m^{-2} K^{-4}.
  Speed of light in vacuo, $\mathbf{c} = 2.99792 \times 10^8$ m s^{-1}.
  Boltzmann's constant, $\mathbf{k} = 1.38066 \times 10^{-23}$ J K^{-1}.
  Planck's constant, $\mathbf{h} = 6.62618 \times 10^{-34}$ J s.
  Atomic mass unit, $\mathbf{u} = 1.66057 \times 10^{-27}$ kg.
  Avagadro's number, $\mathbf{L} = 6.02205 \times 10^{23}$ molecules (g-mol)$^{-1}$.
  Molar gas constant, $\mathbf{R}_m = 8.31441$ J K^{-1} (g-mol)$^{-1}$.

Sun
  Equatorial radius, $R_\odot = 6.9598 \times 10^8$ m.
  Emission temperature, $T_\odot \simeq 5783$ K.
  Mean angular diameter seen from earth, $d\theta_\odot = 31.988$ arc min.
  Mean solid angle seen from earth, $d\omega_\odot = 6.8000 \times 10^{-5}$ sterad.
  Solar constant (1980) = 1373 W m^{-2}.

Earth
  Equatorial radius, $R_e = 6.37816 \times 10^6$ m.
  Mean earth-sun distance (AU), $D_e = 1.4960 \times 10^{11}$ m.
  Eccentricity of orbit, $e = 0.016750$.
  Inclination of rotation axis, $i = 23.45$ deg.
  Standard surface gravity, $g = 9.80665$ m s^{-2}.
  Mass, $M_e = 5.976 \times 10^{24}$ kg.

Atmosphere (dry)

Standard surface pressure, $p_s = 1.01325 \times 10^5$ Pa.

Loschmidt's number, $n_s = 2.688 \times 10^{+25}$ molecules m^{-3}, at stp.

Density at stp, $\rho_s = 1.2930$ kg m^{-3}.

Specific gas constant, $r_a = 2.8706 \times 10^2$ J K^{-1} kg^{-1}.

Mean molecular mass, $\mathcal{M}_a = 28.965$ amu.

Specific heats at stp:

$$c_p = 1.006 \times 10^3 \text{ J kg}^{-1} \text{ K}^{-1},$$
$$c_v = 7.18 \times 10^2 \text{ J kg}^{-1} \text{ K}^{-1},$$
$$\gamma = \tfrac{c_p}{c_v} = 1.401.$$

Water vapor

See Appendix E.

Units

MKS units are used except where usage suggests otherwise.

Pressure, p: The MKS unit of pressure is the Pascal (Pa, N m^{-2}), but meteorologists commonly use the bar (10^5 Pa) and the millibar (mb, 10^2 Pa).

Mixing ratio: Volume mixing ratios, or molecular mixing ratios (μ), are given in percent (%), parts per million (ppmv), parts per billion (ppbv), or parts per trillion (pptv). Mass mixing ratios (m) are given in kilograms per kilogram or grams per kilogram.

Wavelength, λ: No single unit is universal. Common units are nanometers (nm, 10^{-9} m), micrometers (μm, 10^{-6} m), and, occasionally, Ångstrøm units (Å, 10^{-8} m).

Frequency, ν: Inverse centimeters or wavenumbers (cm^{-1}).

Reading

Kaye, G.W.C., and Laby, T.H., 1973, *Tables of physical and chemical constants and some mathematical functions*, 14th ed. London: Longman.

Allen, C.W., 1973, *Astrophysical quantities*. London: Athlone Press.

Appendix B

THERMAL STATE OF THE ATMOSPHERE

The U.S. Standard Atmosphere (Table B.1) is a widely used reference atmosphere that corresponds approximately to a global and annual mean of observed temperatures.

Table B.1 The U.S. Standard Atmosphere

Altitude km	Temperature, K	Pressure Pa	Scale height, km	Molecular density, m^{-3}
0	288.15	1.013×10^5	8.43	2.547×10^{25}
4	262.17	6.166×10^4	7.68	1.704×10^{25}
8	236.21	3.565×10^4	6.93	1.093×10^{25}
12	216.65	1.940×10^4	6.37	6.486×10^{24}
16	216.65	1.035×10^4	6.37	3.461×10^{24}
20	216.65	5.529×10^3	6.38	1.849×10^{24}
24	220.56	2.972×10^3	6.50	9.759×10^{23}
30	226.51	1.197×10^3	6.69	3.828×10^{23}
40	250.35	2.871×10^2	7.42	8.308×10^{22}
50	270.65	7.978×10^1	8.05	2.135×10^{22}
60	247.02	2.196×10^1	7.37	6.439×10^{21}
80	198.64	1.052×10^0	5.96	3.838×10^{20}
100	195.08	3.201×10^{-2}	6.01	1.189×10^{19}
140	559.63	7.203×10^{-2}	20.03	9.322×10^{16}
200	854.56	8.474×10^{-5}	36.19	7.182×10^{15}
500	999.24	3.024×10^{-7}	68.79	2.192×10^{13}
800	999.99	1.704×10^{-8}	193.86	5.442×10^{12}

Figure B.1 shows a time-and-longitude mean of the observed temperatures in the lower and middle atmospheres for the northern winter solstice.

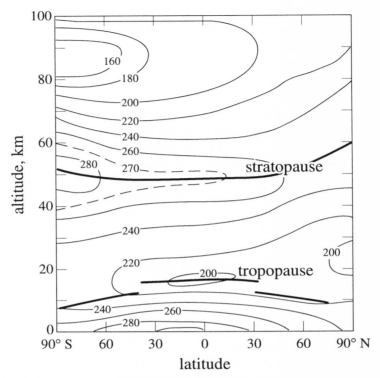

Figure B.1 Observed temperatures of the lower and middle atmosphere at northern winter solstice.

To a first approximation, the two poles may be interchanged to give the temperatures for the northern summer solstice. More detail for the lowest 30 km, separately for each season, is shown in Figures B.2 and B.3.

Reading

Newell, R.E., Kidson, J.W., Vincent, D.G., and Boer, J.G., 1972, *The general circulation of the tropical atmosphere and interactions with extratropical latitudes*, **Vol. 1**. Cambridge, MA: M.I.T. Press.

U.S. Standard Atmosphere, 1976, Publication NOAA-S/T76-1562. Washington, D.C.: U.S. Government Printing Office.

THERMAL STATE OF THE ATMOSPHERE

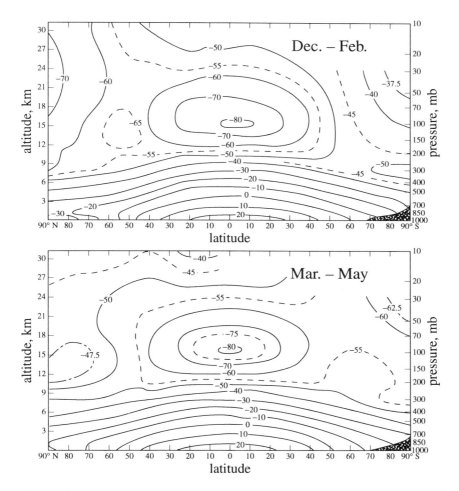

Figure B.2 Temperature of the lower atmosphere in northern winter and spring. These are zonal means of observed data. The isotherms are in °C.

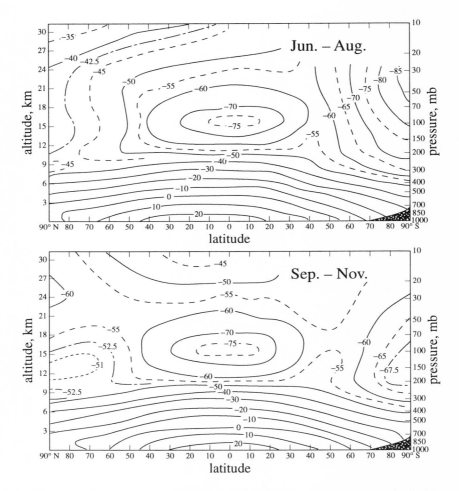

Figure B.3 Temperature of the lower atmosphere in northern summer and fall. See the caption to Figure B.2.

Appendix C

ATMOSPHERIC MOTIONS

Figures C.1 and C.2 give information on the time-and-longitude averages of the three atmospheric wind components. East-west, or *zonal* winds, $\overline{u}$, have magnitudes up to tens of m s^{-1}; north-south, or *meridional* winds, $\overline{v}$, have magnitudes up to 1 m s^{-1}; vertical winds, $\overline{w}$, are measured in mm s^{-1}. Vertical winds cannot be measured directly and must be inferred from the equation of continuity. In terms of the streamfunction, ψ, meridional and vertical winds are given by

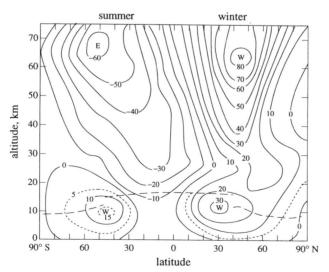

Figure C.1 Mean zonal winds in the middle and lower atmospheres. W and E stand for westerly and easterly, respectively. Meteorological convention names winds after the direction *from which* they come. Westerly is the direction of earth's rotation and is usually taken to be positive. Wind speeds are in m s^{-1}. The dashed line indicates the tropopause.

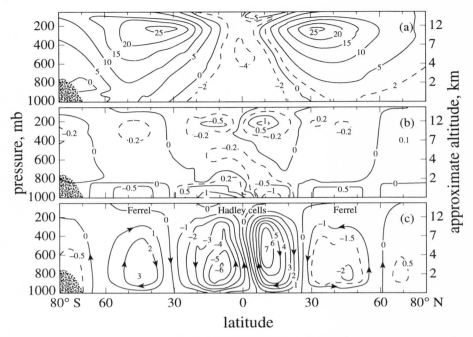

Figure C.2 Time-and-longitude average tropospheric winds. (a) Zonal wind, m s^{-1}; (b) meridional wind, m s^{-1}; (c) mass streamfunction, 10^{10} kg s^{-1}.

$$\overline{v} = \frac{g}{2\pi R_e \cos\phi} \frac{\partial \psi}{\partial p} \text{ (northward, +ve)}, \tag{C.1}$$

$$\overline{w} = \frac{Hg}{p 2\pi R_e^2 \cos\phi} \frac{\partial \psi}{\partial \phi} \text{ (upward, +ve)}. \tag{C.2}$$

g is gravity, p is pressure, R_e is the earth's radius, ϕ is latitude in radians (northward, +ve), and H is scale height. From equations (C.1) and (C.2), orders of magnitude are related by

$$\frac{\overline{w}}{\overline{v}} \sim \frac{H}{R_e} \simeq 1.2 \times 10^{-3}. \tag{C.3}$$

Departures from time-and-longitude means are, for some purposes, treated as large-scale, eddy mixing, with empirical eddy diffusion coefficients (see Chapter 9). The eddy diffusion coefficient, K^e, relates two vectors and is, therefore, a tensor quantity. Little use has been made of other than the two diagonal elements, K_z^e and K_y^e. Some data that have been used in chemical calculations are given in Figures C.3 and C.4.

ATMOSPHERIC MOTIONS

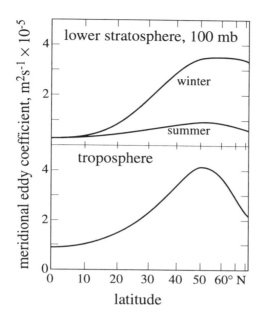

Figure C.3 Meridional eddy diffusion coefficients in the Northern Hemisphere.

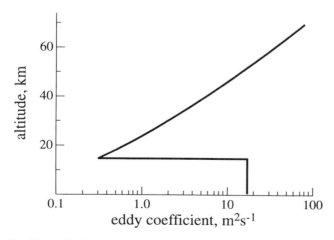

Figure C.4 Vertical eddy diffusion coefficients derived from trace gas measurements.

Reading

Good references include Lindzen (1990), and Peixoto and Oort (1992) in Chapter 2, and Warneck (1988) in Chapter 1. Other references are:

Andrews, D.G., Holton, J.R., and Leovy, C.B., 1987, *Middle atmosphere dynamics*. New York: Academic Press;

Lorenz, E.N., 1967, *The nature and theory of the general circulation of the atmosphere*. Geneva: World Meteorological Organization;

Jursa, A.S. (ed.), 1985, *Handbook of geophysics and the space environment*. National Technical Information Center, ADA 167000.

Appendix D

SOLAR RADIATION

See Appendix A for numerical magnitudes.

The solar spectrum is crossed by absorption lines (*Fraunhofer lines*) throughout the visible spectrum and most of the ultraviolet spectrum. Below 185 nm lines are seen in emission rather than absorption. Both emission and absorption lines are smoothed out in Figure D.1.

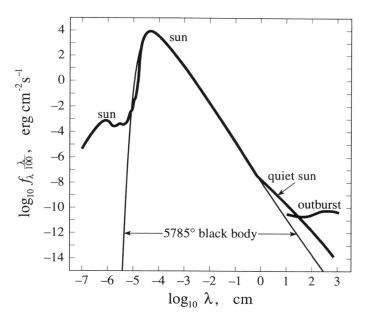

Figure D.1 The solar spectral irradiance from 1 nm to 10 m. All spectral features have been smoothed out. "Outburst" indicates a radio disturbance.

Emission in the ultraviolet spectrum is very variable. The Lyman-α line of hydrogen at 121.6 nm is interesting because it corresponds to a

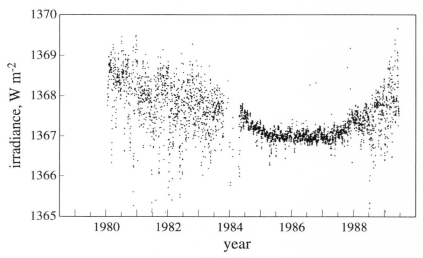

Figure D.2 The solar irradiance integrated from 200 nm to 10 µm. These are direct measurements made outside the atmosphere by R.C. Willson.

narrow window in the absorption spectrum of atmospheric oxygen, and it can influence the atmosphere at lower levels than can most of the actinic ultraviolet radiation from the sun.

99% of the solar energy lies in the spectral region from 200 nm to 10 µm. Measurements of the total radiation in this spectral region, from 1980 to 1989, are shown in Figure D.2. Day-to-day variations are associated with short-lived solar features, such as sunspots and faculae; there is a long-term trend in both the mean and the variability that is associated with the 11.2-y sunspot cycle, a minimum of which occurred in 1986.

The 1980 mean value of the solar irradiance, averaged over all observers, is 1373 W m^{-2}, corresponding to a black-body emission temperature of

$$T_\odot = 5783 \text{ K}.$$

Reading

White, O. (ed.), 1977, *The solar output and its variations*. Boulder, CO: Colorado Associated University Press.

Willson, R.C., 1984, "Measurements of solar total irradiance and its variability," *Space Science Reviews* **38**, 203.

Appendix E

WATER

Numerical data

Molecular mass, $\mathcal{M}_w = 18.015$ amu.
Density of vapor at stp, $\rho_w = 8.031 \times 10^{-1}$ kg m^{-3}.
Specific gas constant, $r_w = 4.615 \times 10^2$ J K^{-1} kg^{-1}.
Latent heat of vaporization of water at 0°C, $l_v = 2.501 \times 10^6$ J kg^{-1}.
Latent heat of fusion of ice at 0°C, $l_f = 3.33 \times 10^5$ J kg^{-1}.
Specific heat of liquid water at 0°C, $c = 4.217 \times 10^3$ J K^{-1} kg^{-1}.
Specific heat of vapor at constant pressure at 298 K, $c_p = 1.864 \times 10^3$ J K^{-1}kg^{-1}.
$\gamma = \frac{c_p}{c_v} = 1.334$ at 373 K.

Vapor pressure

To interpolate in Table E.1, use the approximate integral to the Clausius-Clapeyron equation (G.29),

$$\Delta \log p \propto -\Delta \frac{1}{T}. \tag{E.1}$$

Relative humidities as low as 10% are encountered occasionally at ground level, and 100% humidity occurs wherever there is cloud or mist. On average, the relative humidity of the troposphere in middle latitudes is close to 50% (see Figure E.1).

The rapid variation of saturated vapor pressure with temperature means that most atmospheric water is in the lower atmosphere. The partial pressure of water vapor in the upper troposphere is very small. The e-folding height for the water-vapor partial pressure is about 2 km in the troposphere.

In the middle atmosphere, the water-vapor pressure is determined by complex processes, including photochemistry. Figure E.2 shows that the mixing ratio does not vary greatly above about 12 km (100 mb pressure),

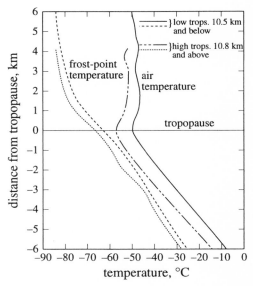

Figure E.1 Average temperature and frost-point temperature measured over southern England. The frost-point temperature is the temperature at which ice condenses from the vapor phase. The water vapor partial pressure can be obtained by entering Table E.1 with the frost-point temperature.

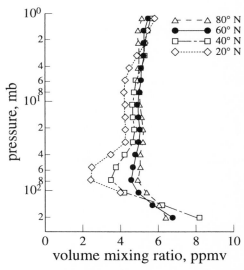

Figure E.2 Water-vapor mixing ratios in the middle atmosphere.

Table E.1 Saturated vapor pressure over water and ice, mb

°C	Over water	Over ice	°C	Over ice
+45	9.586×10^1		−35	2.233×10^{-1}
+40	7.378×10^1		−40	1.283×10^{-1}
+35	5.624×10^1		−45	7.198×10^{-2}
+30	4.243×10^1		−50	3.935×10^{-2}
+25	3.167×10^1		−55	2.092×10^{-2}
+20	2.337×10^1		−60	1.080×10^{-2}
+15	1.704×10^1		−65	5.406×10^{-3}
+10	1.227×10^1		−70	2.618×10^{-3}
+5	8.719×10^0		−75	1.220×10^{-3}
0	6.108×10^0	6.108×10^0	−80	5.472×10^{-4}
−5	4.215×10^0	4.015×10^0	−85	2.353×10^{-4}
−10	2.863×10^0	2.597×10^0	−90	9.672×10^{-5}
−15	1.912×10^0	1.652×10^0	−95	3.784×10^{-5}
−20	1.254×10^0	1.032×10^0	−100	1.403×10^{-5}
−25	8.070×10^{-1}	6.323×10^{-1}		
−30	5.088×10^{-1}	3.798×10^{-1}		

but a slight increase of mixing ratio with height is evident at all latitudes.

Appendix F

THE FLUID EQUATIONS

Continuity

$$\frac{d\rho}{dt} + \rho \nabla \cdot \vec{u} = 0, \tag{F.1}$$

where ρ is density, t is time, and $\vec{u}$ is the vector velocity, with components (u,v,w). The substantial derivative denotes changes moving with the fluid,

$$\frac{d}{dt} \equiv \frac{\partial}{\partial t} + u_i \frac{\partial}{\partial x_i}, \tag{F.2}$$

where $x_i = (x, y, z)$ and $u_i = (u, v, w)$.

An alternate form for equation (F.1) is

$$\frac{\partial \rho}{\partial t} + \nabla \cdot \rho \vec{u} = 0. \tag{F.3}$$

Momentum

In an inertial frame of reference, Newton's second law of motion for a fluid may be expressed in the form,

$$\rho \frac{du_i}{dt} = -\rho g_i + \frac{\partial \sigma_{ij}}{\partial x_j}, \tag{F.4}$$

where σ_{ij} is the *stress tensor* or force in the i-direction acting on unit area of a surface in the j-direction, and $g_i = (0, 0, g)$. σ_{ij} is also the negative of the flux of i-momentum in the j-direction (see Equation F.21 for the definition of flux).

σ_{ij} has a viscous component, $\tau_{ij}(i \neq j)$, and a nonviscous component, $-p$, where p is the pressure,

$$\sigma_{ij} = -p\delta_{ij} + \tau_{ij}. \tag{F.5}$$

The derivative of the viscous component is often expressed in the approximate form

$$\frac{\partial \tau_{ij}}{\partial x_j} = \rho f_i \approx \frac{\partial}{\partial x_j} \rho \nu \frac{\partial u_i}{\partial x_j}, \tag{F.6}$$

where ν is the kinematic viscosity.

The fluid equations are usually transformed to a coordinate system fixed on the earth's surface. This introduces two types of imaginary forces: *centrifugal forces* involving products of velocity components, and *Coriolis forces* involving products of velocities and the angular velocity of the earth. Except in the upper atmosphere, $\frac{z}{R_e} \ll 1$, where R_e is the earth's radius, and may be neglected. With this approximation the momentum equations may be written

$$\frac{du}{dt} - \frac{uv \tan \phi}{R_e} + \frac{uw}{R_e} = -\frac{1}{\rho}\frac{\partial p}{\partial x} + 2\Omega v \sin \phi - 2\Omega w \cos \phi + f_x, \tag{F.7}$$

$$\frac{dv}{dt} + \frac{u^2 \tan \phi}{R_e} + \frac{vw}{R_e} = -\frac{1}{\rho}\frac{\partial p}{\partial y} - 2\Omega u \sin \phi + f_y, \tag{F.8}$$

$$\frac{dw}{dt} - \frac{u^2 + v^2}{R_e} = -\frac{1}{\rho}\frac{\partial p}{\partial z} - g + 2\Omega u \cos \phi + f_z. \tag{F.9}$$

ϕ is the latitude, Ω is the earth's rotational speed, and (f_x, f_y, f_z) are the frictional body forces.

Geostrophic and thermal wind equations

On the basis of the orders of magnitude of the terms in (F.7), (F.8), and (F.9) for a typical tropospheric, synoptic system, the equations may be approximated,

$$\frac{du}{dt} \approx -\frac{1}{\rho}\frac{\partial p}{\partial x} + 2\Omega v \sin \phi + f_x, \tag{F.10}$$

$$\frac{dv}{dt} \approx -\frac{1}{\rho}\frac{\partial p}{\partial y} - 2\Omega u \sin \phi + f_y, \tag{F.11}$$

$$\frac{dw}{dt} \approx -\frac{1}{\rho}\frac{\partial p}{\partial z} - g \approx 0. \tag{F.12}$$

The approximations (F.10), (F.11), and (F.12) may have to be revised for circumstances when velocities are large, such as in a convective cloud. Equation (F.12) is the *hydrostatic approximation*. For quasi-one-dimensional motions, (F.12) may be written

$$dp = -\rho g dz. \tag{F.13}$$

THE FLUID EQUATIONS

For planetary-scale, steady-state motions, equations (F.10) and (F.11) are usually further approximated to

$$0 \approx -\frac{1}{\rho}\frac{\partial p}{\partial x} + 2\Omega v \sin\phi, \tag{F.14}$$

$$0 \approx -\frac{1}{\rho}\frac{\partial p}{\partial y} - 2\Omega u \sin\phi. \tag{F.15}$$

(F.14) and (F.15) together constitute the *geostrophic approximation*. They describe motions that are orthogonal to pressure gradients.

With slight approximation to the boundary condition at the lower surface, the geostrophic equations and the hydrostatic approximation yield the *thermal wind* equations:

$$\frac{\partial u}{\partial z} \approx -\frac{g}{2\Omega T \sin\phi}\frac{\partial T}{\partial y}, \tag{F.16}$$

$$\frac{\partial v}{\partial z} \approx \frac{g}{2\Omega T \sin\phi}\frac{\partial T}{\partial x}. \tag{F.17}$$

Useful theorems

Differentiating integrals

A volume of fluid, $V(t)$, will be distorted by the flow and it may be shown that

$$\frac{d}{dt}\int_V f\rho dV = \int_V \rho\frac{df}{dt}dV. \tag{F.18}$$

Gauss' theorem

$$\int_V \nabla\cdot\vec{F}dV = \int_A F_n dA, \tag{F.19}$$

where A is the surface surrounding V and F_n is the *outward* normal component[1] of the fluid flux $\vec{F}$.

Flux divergence

$$\rho\frac{d\xi}{dt} \equiv \nabla\cdot\vec{F}(\xi) + \frac{\partial(\rho\xi)}{\partial t}, \tag{F.20}$$

where

$$\vec{F}(\xi) = \vec{u}\rho\xi \tag{F.21}$$

is the flux of the state function, ξ.

Reading

Lamb, H., 1945, *Hydrodynamics*. New York: Dover Publications, and Lindzen (1990), Chapter 2.

---

[1] In the atmosphere it is usual to define flux components to be *positive upward*, leading to ambiguities in some cases. For example, at the earth's surface, the vertical flux component in the atmosphere, F_z, and the outward component, F_n, have opposite signs.

Appendix G

THERMODYNAMIC CONCEPTS

The essence of classical thermodynamics is to identify bulk properties of matter, or *state functions*, that are uniquely determined when the state is defined, and to use them for bookkeeping purposes. The two most important state functions are *internal energy* and *entropy*; *enthalpy*, the *Gibbs function*, and *availability* are derived functions, useful for special purposes. *Heat* and *work* are not state functions but are *local interactions* that modify state functions.

Intensive variables

T	temperature
p	pressure
m	mixing ratio

Extensive variables

v, V	volume	$v = \rho^{-1}$
q, Q	heat interaction	
ϕ, Φ	work interaction	
e, E	internal energy	
h, H	enthalpy	$h = e + pv$
s, S	entropy	
a, A	availability	$a = e - T_s s$
g, G	Gibbs function	$g = e - Ts + pv$
r, R	gas constant	
c, C	specific heat	

The lower case refers to unit mass (a *specific variable*), while the upper case refers to a complete system. T_s is the temperature of the surroundings.

The first law
$$de = dq + d\phi. \tag{G.1}$$

dq and $d\phi$ are incomplete differentials because q and ϕ are not state functions. The implication of this statement is that heat and work cannot be mixed and advected by motions as are the state functions e, h, s, etc. Thus, it is not permissible to talk of a *fluid flux of heat* in the same way as a fluid flux of a state variable (see Equation F.21).

Reversible work that affects the thermal state is

$$d\phi_{\text{rev}} = -pdv. \tag{G.2}$$

Note that for mechanical work to change the thermodynamic state of the system it is necessary to change the volume. Reversible work that does not distort the system changes either the kinetic or the potential energy.

The second law

$$ds \geq \frac{dq}{T}. \tag{G.3}$$

If the changes are reversible, as they will be if there is a unique temperature, and if the work is described by Equation (G.2), the equality applies.

Closed systems

For closed systems, the following are the differential relations between state functions:

$$de = Tds - pdv, \tag{G.4}$$
$$dh = Tds + vdp, \tag{G.5}$$
$$dg = -sdT + vdp, \tag{G.6}$$
$$da = de - T_s ds. \tag{G.7}$$

Open systems

A phase is an open system to which species may be added or subtracted when phase changes occur. To accommodate this possibility we add to Equations (G.4) to (G.7) the same extra term involving the partial chemical potential, g_k^ϕ, where k represents a chemical component and ϕ represents a phase. In the lower atmosphere, the only significant variable component is water. For a single phase, the differential relation for the Gibbs function becomes

$$dG^\phi = -S^\phi dT + V^\phi dp + \sum_{k=1}^{c} g_k^\phi dM_k^\phi. \tag{G.8}$$

dM_k^ϕ is the mass of a species entering a phase.

For a surface phase, the term gdM in Equation (G.8) should be replaced by $-\sigma dA$, where A is the area of the surface and σ is the surface tension.

THERMODYNAMIC CONCEPTS

Condition for equilibrium

For a system at constant pressure and constant temperature,

$$dG = 0. \tag{G.9}$$

G is a *minimum*.

Ideal gases

For practical purposes, air, even with an admixture of water vapor, may be treated as an ideal gas.

Equation of state

$$pv = rT, \tag{G.10}$$

where r is the gas constant (an extensive variable) or

$$p = n\mathbf{k}T, \tag{G.11}$$

where n is the molecular number density and $\mathbf{k}$ is Boltzmann's constant.

Dalton's law for ideal gas mixtures

The pressure, and all extensive thermodynamic functions, obey addition laws of the kind

$$p = \sum_i p_i, \tag{G.12}$$

$$S = \sum_i S_i, \tag{G.13}$$

where the suffixes indicate different molecular species, and p_i and S_i are the pressure and entropy that one species would have if it alone occupied the entire system. If an extensive variable, including the gas constant and the specific heat, has values ξ_a and ξ_v for air and water vapor, respectively, the state function of a mixture is, according to Dalton's law,

$$\xi = (1 - m_v)\xi_a + m_v \xi_v, \tag{G.14}$$

where m_v is the mass mixing ratio of the vapor.

In the absence of phase changes:[1]

$$e = c_v T + \text{const.}, \tag{G.15}$$

where c_v is the specific heat at constant volume;

$$h = c_p T + \text{const.}, \tag{G.16}$$

---

[1] A purist may object to expressions such as $\ln p$ in some of the following equations, but because only differences of entropy and Gibbs function are under consideration here, the units of pressure and temperature are unimportant.

where c_p is the specific heat at constant pressure,

$$c_p - c_v = r, \qquad (G.17)$$

$$s(p, T) = s_0 - r \ln p + c_p \ln T, \qquad (G.18)$$

and

$$g(p, T) = g_0(T) + rT \ln p. \qquad (G.19)$$

For a gas mixture in which phase changes are not taking place, the above equations are valid provided that specific heats and gas constants are calculated for the mixture from Equation (G.14).

Phase changes

All that follows in this section is valid only if two phases, in this instance water vapor (subscript v) and a condensed phase (subscript c), are in equilibrium at the same temperature and pressure. Under this circumstance the atmosphere has only one degree of freedom, and vapor pressure and all state functions of the vapor are functions of temperature only. To emphasize this we write the vapor pressure as $p_e(T)$ in place of p_v. At the transition between phases

$$s_v(T) = s_c(T) + \frac{l(T)}{T}, \qquad (G.20)$$

$$h_v(T) = h_c(T) + l(T), \qquad (G.21)$$

$$e_v(T) = e_c(T) + l(T) - p_e(T)(v_v - v_c), \qquad (G.22)$$

where $l(T)$ is the latent heat for the transition. It is normal to neglect v_c in comparison with v_v, and $p_e(T)v_v$ in comparison with $l(T)$, so that

$$e_v(T) \approx e_c(T) + l(T). \qquad (G.23)$$

For temperature and pressure changes in the vapor phase, use Equations (G.16), (G.17), and (G.19) with r and c_p appropriate to the gas mixture. For temperature and pressure changes in the condensed phase, it is usual to assume incompressibility and a single specific heat, c. Hence,

$$e_c(T) = h_c(T) = cT + \text{const.}, \qquad (G.24)$$

$$s_c(T) = c \ln T + \text{const.} \qquad (G.25)$$

Clausius-Clapeyron equation

At the transition between vapor and condensed phases, the equilibrium vapor pressure $p_e(T)$ is governed by

$$\frac{dp_e(T)}{dT} = \frac{l(T)}{T(v_v - v_c)} \approx \frac{l(T)}{Tv_v} \approx \frac{p_e(T)l(T)}{r_v T^2}. \qquad (G.26)$$

THERMODYNAMIC CONCEPTS 311

Kirchoff's relation

$$l(T) = l(T_0) + (c_{p,v} - c)(T - T_0). \tag{G.27}$$

If the specific heats are constant, Equations (G.8), (G.22), and (G.23) may be combined and integrated to yield

$$\ln p_e(T) = -\frac{\alpha}{T} + \beta \ln T + \gamma, \tag{G.28}$$

where α, β, and γ are constants. The data in Appendix E may be fitted with high precision to equation (G.28). For small departures from a standard temperature, T_0,

$$\ln \frac{p_e(T)}{p_e(T_0)} \approx -\frac{l(T_0)}{r_v} \left(\frac{1}{T} - \frac{1}{T_0} \right). \tag{G.29}$$

Ideal solutions

Raoult's law

$$\frac{p_e(x_l, T)}{p_e(x_l = 1, T)} = x_l = \frac{n_l}{n_l + \nu n_s}, \tag{G.30}$$

where n_l is the number density of liquid molecules, n_s is the number density of solute molecules, and ν is the number of ionized components of the solute.

Reading

Zemansky, M.W., and Dittman, R.H., 1979, *Heat and thermodynamics*, 6th ed. New York: McGraw-Hill.

Appendix H

KINETIC RATES AND RATE COEFFICIENTS

The evolution of chemical species following photolysis is governed by the laws of *stoichiometry* and *mass action*. Stoichiometry recognizes that in chemical reactions atoms are indestructible and that only rearrangements of atoms among molecules will take place. Mass action requires that reaction rates are proportional to collision frequencies and, therefore, to concentrations raised to the powers of molar amounts. For example, the reaction

$$A + 2B \rightarrow \text{products}, \tag{H.1}$$

where A and B are molecules, has, according to the law of mass action, the rate

$$\frac{d[A]}{dt} = \frac{1}{2}\frac{d[B]}{dt} = -k_{AB}[A][B]^2, \tag{H.2}$$

where [A] denotes the molecular number density of A. The *rate constant*, k_{AB}, is a function of temperature only. Rate constants are available for hundreds of important atmospheric reactions, and the number increases as new problems arise.

Equation (H.1) is a *third order* or *trimolecular reaction*. Processes 2 and 4 in Table 1.5 are *second order* or *bimolecular reactions*. Reactions 1 and 3 are *first order* or *monomolecular reactions*. Mass action applied to a monomolecular reaction leads to

$$\frac{d[A]}{dt} = -k_A[A], \tag{H.3}$$

which may be integrated to

$$[A] = [A]_0 \exp-\frac{t}{\tau}, \tag{H.4}$$

where

$$\tau = \frac{1}{k_A} \tag{H.5}$$

is the lifetime of the species and $[A]_0$ is the concentration at time $t = 0$. This lifetime is well defined. For reactions of higher order, the lifetime of species A may be written formally

$$\tau_A^{-1} = -\frac{1}{[A]}\frac{d[A]}{dt}. \tag{H.6}$$

For the bimolecular reaction A+B → products, this definition gives

$$\tau_A^{-1} = k_{AB}[B]. \tag{H.7}$$

If [B] is much greater than [A], the latter can vary while the former is effectively constant, and (H.7) is a useful, working definition of a lifetime.

The magnitudes of rate coefficients are of paramount importance. We shall briefly discuss bimolecular reactions because they are important and relatively simple to understand. For two molecules to react they must first come into contact. The largest possible rate coefficients (*kinetic theory coefficients*) occur if every collision leads to a reaction. From equation (H.7) the rate constant is

$$k(\text{kinetic}) = \frac{1}{\tau_c n}\ \text{m}^3\text{s}^{-1}, \tag{H.8}$$

where τ_c is the time between collisions, and n is the molecular number density. At stp, $n = 2.7 \times 10^{25}$ m^{-3}, $\tau_c \sim 10^{-10}$ s, and $k(\text{kinetic}) \sim 3 \times 10^{-16}$ m^{3}s^{-1}.

Most collisions between reacting species do not result in a reaction. There is usually an energy barrier between reactants and products, the *activation energy* (E_a), which requires an energetic collision to surmount it. The number of collisions with energy greater than E_a involves a Boltzmann factor, and leads to an approximate expression for the rate coefficient known as the *Arrhenius relation*,

$$k(T) = k_0 e^{-\frac{E_a}{r_m T}}, \tag{H.9}$$

where r_m is the molar gas constant and E_a is the energy per mole. The coefficient k_0 is less than or equal to the kinetic theory rate constant. If $E_a \gg r_m T$, (H.9) can lead to a very rapid variation of rate constant with temperature.

Quantum mechanical considerations also enter. The reactants may be regarded as a large molecule in a given quantum state that makes a transition to another state, representing the products. Quantum transitions are subject to more-or-less rigorous selection rules, one of the strictest being that the total electronic spin should not change. Spin forbidden reactions, in which the electronic spin is required to change, are slow.

Reading

See Warneck (1988) and Brasseur and Solomon (1984), Chapter 1.

Appendix I

CLOUD GENERA

Cirrus. Detached clouds in the form of white, delicate filaments or white or mostly white patches or narrow bands. These clouds have a fibrous (hairlike) appearance, or a silky sheen, or both.

Cirrocumulus. Thin, white patch, sheet or layer of cloud without shading, composed of very small elements in the form of grains, ripples, etc. merged or separate, and more or less regularly arranged; most of the elements have an apparent width less than 1 deg.

Cirrostratus. Transparent, whitish cloud veil of fibrous (hairlike) or smooth appearance, totally or partly covering the sky, and generally producing halo phenomena.

Altocumulus. White or grey, or both white and grey, patch, sheet, or layer of cloud, generally with shading, composed of laminae, rounded masses, rolls, etc., which are sometimes partly fibrous or diffuse and which may or may not be merged; most of the regularly arranged small elements usually have an apparent width between 1 and 5 deg.

Altostratus. Grayish or bluish cloud sheet or layer of striated, fibrous, or uniform appearance, totally or partly covering the sky, and having parts thin enough to reveal the sun at least vaguely, as through ground glass. Altostratus does not show halo phenomena.

Nimbostratus. Grey cloud layer, often dark, the appearance of which is rendered diffuse by more or less continuously falling rain or snow, which in most cases reaches the ground. It is thick enough to blot out the sun. Low, ragged clouds frequently occur below the layer, with which they may or may not merge.

Stratocumulus. Grey or whitish, or both grey and whitish, patch, sheet, or layer of cloud, which almost always has dark parts, composed of tessellations, rounded masses, rolls, etc., which are nonfibrous (except for virga) and which may or may not be merged; most of the regularly arranged small elements have an apparent width of more than 5 deg.

Stratus. Generally grey cloud with a fairly uniform base, which may give drizzle, ice prisms, or snow grains. When the sun is visible through the clouds, its outline is clearly discernible. Stratus does not produce halo phenomena, except, possibly, at very low temperatures. Sometimes stratus appears in the form of ragged patches.

Cumulus. Detached clouds, generally dense and with sharp outlines, developing vertically in the form of rising mounds, domes, or towers, of which the bulging upper part often resembles a cauliflower. The sunlit parts of these clouds are mostly brilliant white; their base is relatively dark and nearly horizontal. Sometimes cumulus is ragged.

Cumulonimbus. Heavy and dense clouds, with a considerable vertical extent, in the form of a mountain or huge towers. At least part of its upper portion is smooth, or fibrous or striated, and nearly always flattened; this part often spreads out in the shape of an anvil or vast plume. Under the base of this cloud, which is often very dark, there are frequently low ragged clouds, either merged with it or not, and precipitation sometimes in the form of virga.

Reading

The above descriptions are taken verbatim from:

World Meteorological Organization, 1956, *International Cloud Atlas* Volumes I and II. Geneva, Switzerland: WMO.

Appendix J

PROPERTIES OF SPHERICAL DROPS

From *Stoke's law*, the terminal fall speed of a droplet of radius r is

$$w = \frac{2\rho_l g r^2}{9\mu} = 1.2 \times 10^8 r^2 \text{ (at 1013 mb, and 20°C)}, \tag{J.1}$$

where μ is the coefficient of viscosity. The constant is correct for MKS units only.

Equation (J.1) is accurate for drops of radius less than 20 μm; fall speeds for larger drops are given in Table J.1. Collision efficiencies between drops of radii r_1 and r_2 are given in Figure J.1.

Reading

See Mason (1971) and Pruppacher and Klett (1978), Chapter 8.

Table **J.1** Terminal fall speeds for water drops at 1013 mb pressure, and 20°C

Diameter mm	Speed cm s^{-1}
0.01	0.3
0.03	2.6
0.10	25.6
0.30	115
1.0	403
2.0	649
3.0	806
4.0	883
5.0	909

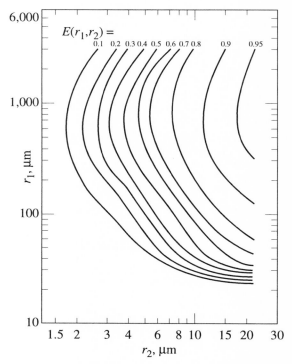

Figure J.1 Collision efficiencies, $E(r_1, r_2)$. r_1 is the larger of the two radii.

INDEX OF DEFINITIONS

Absorption, 16, 58
   coefficient, 59
   efficiency, 76
Acid rain, 196
Activation energy, 16, 314
Activity spectrum, 231
Adding method, 105
Ageostrophic drive, 28
Adiabatic water content, 223
Adjustment models (6.5 K km^{-1}), 129
Aerodynamically rough flow, 225
Aerological diagram, 50
Aerosols, 228
Aitken nuclei, 229
Amplification factor (feedback), 145
Anomalous diffraction, 72
Anvil, 243
Arrhenius relation, 314
Asymmetry factor, 114
Aureole, 82
Availability, 47
Available potential energy, 47

Backward scattering, 114
Band strengths, 122
Biosphere, 5
Black-body radiation, 93
Boltzmann's law, 12
Boussinesq approximation, 258
Bowen ratio, 276
Bright band, 251
Brightness temperature, 148
Bunch clouds, 221
Buoyancy, 260

Capping inversion, 270
Carbonaceous chondrites, 3
Catalyst, 19
Cavity radiation, 93
Centrifugal forces, 304
Chain-terminating step, 19
Chapman
   theory (of ozone), 161
   layer, 135
Chemical time constant, 186
Cirrocumulus, 221
Cirrus, 221
Climate sensitivities, 142
Closure techniques, 263
Cloud
   condensation nuclei, 231
   genera, 221
CO_2-slicing, 251
Coagulation, 237
Collision cross section, 237
Convective boundary layer, 243
Condensation
   nuclei, 223
   temperature, 29
Conditionally unstable, 35
Constant-stress layer, 256, 263
Continuous growth, 237
Continuum
   absorption, 59
   contributions, 69
Convective boundary layer, 270
Coriolis forces, 304
Correlated-k method, 110
Cosmic abundances, 2
Critical
   level, 221

319

Richardson number, 267
Cumulative
    k-distribution, 109
    number density, 230
Cumulonimbus clouds, 243
Cumulus
    clouds, 221
    convection model, 131
    stage (in cumulonimbus), 243
Deep ocean, 208
Density scale height, 266
Destruction (of ozone), 177
Detrainment level, 132
Dielectric spheres, 80
Diffraction, 77
    peak, 77
    rings, 77
Diffusive separation, 12
Dissipating stage (of cumulonimbus), 243
Dissipation, 25
Dissociation, 15
Doppler line
    shape, 65
    width, 65
Doubling
    -and-adding method, 102
    method, 105
Dry
    adiabatic lapse rate, 33
    air, 12
    deposition, 199
    potential temperature, 30
Drizzle, 245

Eddies, 257
Eddy
    diffusion coefficient, 263, 264
    diffusion time, 266
    energy, 268
    flux, 257
    stresses, 257
Ekman
    layer, 262
    spiral, 262
Electronic transitions, 60
Elsasser band model, 110
Emission
    level, 135
    temperature, 101, 135
Energy
    balance models, 139
    intensity (radiance), 90
Enthalpy, 26, 307
Entrainment fluxes, 271
Entropy, 27, 307
Equilibrium
    radiation, 93
    systems, 8
Equivalent
    potential temperature, 30
    width, 84
Escape
    level, 21
    velocity, 4
Excess sulfate, 197
Extinction, 58
    coefficient, 58
    efficiency, 76
Exobase, 21
Exosphere, 4

First-order (monomolecular) reaction, 313
Fluid flux of heat, 308
Flux Richardson number, 266
Fluxes (between reservoirs), 204
Form drag, 255
Forward scattering, 114
Fraunhofer lines, 297
Free radicals, 16
Freezing nuclei, 232
Frequency-coherent process, 98
Friction velocity

Gas-to-particle conversion, 231
Geopotential height, 26, 177
Geostrophic approximation, 305
Gibbs function, 307

Giant nuclei, 229
Global
    evaporation rate, 43
    precipitation rate, 43
Greenhouse effect, 130
Grey absorption, 125
Graupel, 245

Hail, 245
Heat, 307
High
    clouds, 221
    clouds (and radiation), 146
Homogeneous nucleation, 227
Homogeneous reactions, 182
Hydrostatic approximation, 304

Ice nuclei, 232
Incoherent scattering, 98
Induced emission, 97
Initial steps, 15
Inorganic (carbon), 206
Interaction principle, 91, 103
Interactive heat flux, 24
Intermediate storage (of carbon), 208
Internal
    cycling, 169
    energy, 26, 307
Invariant imbedding, 121
Inventories, 36
Inverse lifetime (rate), 163
Inversion layer, 244
Ionosphere, 6
Irreversible work, 25
Isotropic scattering, 114

Jeans escape, 4
Junge
    distribution, 230
    layer, 203

K-closure, 262
k-distribution method, 109
Kelvin relation, 226
Kernel function, 136

Kinetic
    energy, 26
    temperature, 13
    theory coefficients, 314

Lambert's law, 58
Laplace's theory, 2
Lapse rate, 33
Large nuclei, 229
Limb
    brightening, 154
    darkening, 154
Line
    -by-line calculations, 107
    contributions, 69
    shape, 65
    -shift parameter, 67
    spectra, 59
    strengths, 61
Liquid-water content, 238
Local
    closure, 263
    dissipation, 261
    thermodynamic equilibrium, 12
Logarithmic profile, 256
Low clouds, 221
    (and radiation), 146
Lower atmosphere, 7

Macrophysical (properties), 221
Mass action, 313
Mature stage (of a cumulonimbus), 243
Mean free path (of a photon), 59
Melting band, 251
Meridional winds, 293
Mesoscale convective systems, 243
Mesosphere, 6
Method of moments, 112
Michelson-Lorentz line
    shape, 66
    width, 66
Microphysical (properties), 221
Middle
    atmosphere, 7

clouds, 221
Mie's theory (of scattering), 70
Mixed layer (of the ocean), 208
Mixing length, 264
Moist-adiabatic
    adjustment model, 131
    lapse rate, 34
Molecular diffusion layer, 42
Monin-Obukhov
    length, 267
    similarity theory, 262
Monochromatic
    quantities, 107
    radiative equilibrium, 92
Multiple scattering, 74
Multiplication mechanisms, 247

Natural decay time, 17
Neutral, 35
    flow, 256, 265
Newtonian approximation, 116
Nonhomogeneous reactions, 182
Nonlocal dissipation, 262

Odd
    halogen compounds, 162
    hydrogen compounds, 162
    oxygen compounds, 162
    nitrogen compounds, 162
Optical
    depth, 75, 100
    path, 58, 99
Organic (carbon), 205
Orographic clouds, 221
Oxidized
    group (sulfur compounds), 196
    species (carbon compounds), 205
Ozone hole, 182

Peroxy free radical, 192
Phase function for scattering, 75
Photochemical
    balance, 179
    loss rate, 163
    process, 16

Photolysis, 16
Planck function, 93
Plane of reference, 72
Planetary
    albedo, 126
    boundary layer, 254
    radiation, 8
Polar stratospheric clouds, 182
Potential
    energy, 26
    temperature, 30
Production (of ozone), 177
Protoplanetary nebula, 2
Purple light, 204

Quenching, 16

Radiant intensity, 58
Radiance, 58, 90
Radiation-to-space approximation, 115
Radiative
    decay, 16
    equilibrium, 92
    transfer theory, 58
Rain, 245
Raoult's law, 226
Rate
    constant, 313
    limiting step, 189
Rayleigh
    -Jeans radiation law, 93
    scattering, 59
Rayleigh's theory (of scattering), 71
Reaction, 16
Reduced
    group (of sulfur compounds), 196
    species (of carbon), 206
Reflection functions, 104
Relative humidity, 227
Remote sensing, 147
Reservoirs (of carbon), 204
Reversible work, 25
Reynold's stress tensor, 259
Rock-forming elements, 2
Rotational transitions, 60

INDEX OF DEFINITIONS

Roughness
  layer, 255
  length, 256
Runaway greenhouse, 145

Saturated adiabatic lapse rate, 34
Scalar polarizability, 72
Scale height, 10
Scattering, 16
  angle, 71, 73
  coefficient, 59
  efficiency, 76
Schwarzschild's equation of transfer, 91
Scotch mist, 248
Second order (bimolecular) reaction, 313
Semi-
  grey model, 124
  isotropic field of radiation, 113
Shear generation, 260
Sheet clouds, 221
Short-term storage (of carbon), 208
Showers, 245
Similarity theories, 266
Single-scattering albedo, 58
Size parameter, 71
Skin temperature, 127
Snow, 245
Solar constant, 96
Source function, 91
  for multiple scattering, 96
  for single scattering, 96
Solar irradiance, 96
Solute, 225
Specific variable, 307
Spectral resolution, 85
Stable (atmosphere), 34
State functions, 307
Static energy, 27
Stefan-Boltzmann law, 94
Stoichiometry, 313
Stratosphere, 6
Stratified atmosphere, 9

Stratocumulus, 221
Status, 221
Stokes parameters, 95
Stress tensor, 303
Strong
  lines, 110
  -line limit, 84
Sublimation, 232
Supersaturation, 227
Surface
  layer, 256, 263
  stress, 255

Tangent height, 157
Tephigram, 50
Terminal fall speeds, 237
Terrestrial radiation, 8
Thermalization, 13
Thermalization time, 16
Thermal
  radiation, 8
  reservoir, 15
  wind, 305
Thermochemical equilibrium, 15
Thermocline, 208
Thermosphere, 6
Third order (trimolecular) reaction, 313
Total potential energy, 27
Transmission, 58
  functions, 104, 108
Transport of eddy energy, 260
Troposphere, 6
Two-stream model, 124

Unstable (atmosphere), 35
Upper atmosphere, 7

Vibrational transitions, 60
Vibration-rotation band, 60
Virtual potential temperature, 30
Viscous
  sublayer, 254
  work, 27
Voigt profile, 67

Volatile compounds, 3
Volume mixing ratio, 6
von Karman's constant, 256

Water-vapor
    continuum, 68
    feedback, 145
Weak
    -line limit, 84
    lines 110
Wegener-Bergeron process, 247
Wet deposition, 199
Wien's
    displacement law, 93
    radiation law, 93
Work, 307

Zenith angle, 95
Zero-point displacement, 256
Zonal winds, 293